The Art of Identification

THE SLSA BOOK SERIES

Lucinda Cole and Robert Markley, General Editors

Advisory Board:
Stacy Alaimo (University of Texas at Arlington)
Ron Broglio (Arizona State University)
Carol Colatrella (Georgia Institute of Technology)
Heidi Hutner (Stony Brook University)
Stephanie LeMenager (University of Oregon)
Christopher Morris (University of Texas at Arlington)
Laura Otis (Emory University)
Will Potter (Washington, DC)
Ronald Schleifer (University of Oklahoma)
Susan Squier (Pennsylvania State University)
Rajani Sudan (Southern Methodist University)
Kari Weil (Wesleyan University)

Published in collaboration with the Society for Literature, Science, and the Arts, AnthropoScene presents books that examine relationships and points of intersection among the natural, biological, and applied sciences and the literary, visual, and performing arts. Books in the series promote new kinds of cross-disciplinary thinking arising from the idea that humans are changing the planet and its environments in radical and irreversible ways.

The Art of Identification

Forensics, Surveillance, Identity

Edited by
Rex Ferguson,
Melissa M. Littlefield,
and James Purdon

The Pennsylvania
State University Press
University Park,
Pennsylvania

Library of Congress Cataloging-in-Publication Data

Names: Ferguson, Rex, 1977– editor. | Littlefield, Melissa M., 1979– editor. | Purdon, James, 1983– editor.
Title: The art of identification : forensics, surveillance, identity / edited by Rex Ferguson, Melissa M. Littlefield, and James Purdon.
Other titles: AnthropoScene.
Description: University Park, Pennsylvania : The Pennsylvania State University Press, [2021] | Series: AnthropoScene: the SLSA book series | Includes bibliographical references and index.
Summary: "A multidisciplinary collection of essays exploring current scholarship on the history of human identification. Examines how techniques of identification are entangled within a wider sphere of cultural identity formation"—Provided by publisher.
Identifiers: LCCN 2021021596 | ISBN 9780271090573 (hardback)
Subjects: LCSH: Identity (Philosophical concept) in literature—History. | Identity (Philosophical concept)—History. | Identification—History. | Identification—Social aspects—History. | LCGFT: Essays.
Classification: LCC PN56.I42 A78 2021 | DDC 111/.82—dc23
LC record available at https://lccn.loc.gov/2021021596

Copyright © 2021 The Pennsylvania State University
All rights reserved
Printed in the United States of America
Published by The Pennsylvania State University Press,
University Park, PA 16802-1003

Photography by Eric Fong

The Pennsylvania State University Press is a member of the Association of University Presses.

It is the policy of The Pennsylvania State University Press to use acid-free paper. Publications on uncoated stock satisfy the minimum requirements of American National Standard for Information Sciences—Permanence of Paper for Printed Library Material, ANSI Z39.48–1992.

Contents

Acknowledgments | vii

Introduction | 1
Rex Ferguson, Melissa M. Littlefield,
and James Purdon

Part 1 | Genres of Identification

1   Charming Faces and the Problem
    of Identification | 23
    Matt Houlbrook

2   Identity Noir | 49
    James Purdon

3   "The Ghosts of Individual Peculiarities":
    Murder and Interpretation in Dickens | 65
    Andrew Mangham

4   "A Puzzle of Character": Francis Iles and
    Narratives of Criminality in the 1930s | 82
    Victoria Stewart

Part 2 | The Body Captured

5   The Art of Identification:
    The Skeleton and Human Identity | 101
    Rebecca Gowland and Tim Thompson

6   Becoming More Biological: Ruth Ozeki
    and the Postgenomic Ethnoracial Novel | 121
    Patricia E. Chu

7   Identification Made Visible: Photographic
    Evidence and Russell Williams | 139
    Jonathan Finn

Part 3 | Surveillant Technologies

8   The Face in the Biometric Passport | 159
    Liv Hausken

9   The Bourne Identification | 182
    Rex Ferguson

10 Identification and the "Intelligent City" | 202
    Dorothy Butchard

11 Jennifer Egan and the Database | 223
    Rob Lederer

Contributors | 245

Index | 249

Acknowledgments

The Art of Identification began life as a series of workshops held at the University of Birmingham, the University of St. Andrews, and the University of Illinois at Urbana-Champaign that could not have taken place without the support of a research networking grant from the Arts and Humanities Research Council. The conversations begun at those workshops shaped the overall design of this volume and informed the majority of its chapters. The editors are therefore indebted to all workshop contributors, including Jess Woodhams, Lee Rainbow, Warren Hines, Dan Vyleta, Richard House, Deborah Jermyn, Jane Caplan, Kamilla Elliott, Spencer Schaffner, Jonathan Finn, Kate West, Charlotte Bilby, Annie Ring, Dawn Stobbart, Simon Cole, Stephen Cartwright, Kenworthey Bilz, Fabienne Collignon, Ian Burney, Malthe Boye Bjerregard, June Jones, and Jakob Stougaard-Nielsen. Meeting the artist Eric Fong (www.ericfong.com) in the final workshop was one of the wonderful discoveries of the series, and we gratefully acknowledge his contribution to the volume of the images found on the front cover and on the introductory pages to parts 1, 2, and 3. In pursuing publication, we were fortunate to have the support of Lucinda Cole and Bob Markley, series editors at Penn State University Press, who championed the project. At the press, our thanks go to Kendra Boileau, Alex Vose, and our anonymous readers, whose perceptive suggestions helped to strengthen the final manuscript in numerous ways.

Introduction

*Rex Ferguson,
Melissa M. Littlefield,
and James Purdon*

Ronald Pinn was born in Bermondsey, London, on 23 January 1964 and died on 9 August 1984, aged twenty years. And he would have remained deceased had British writer Andrew O'Hagan not assumed Pinn's identity during a twenty-first-century experiment in contemporary self-invention. While researching covert practices developed by the London Metropolitan Police beginning in the 1960s, O'Hagan had found that, in the dangerous work of infiltrating extremist groups and passing as members of these groups, undercover Met officers had to become masters at mimicking the authentic behaviors, looks, and language of credible members of the group. But what was equally intrinsic to a successful impersonation was the creation of what the police and intelligence services refer to as a "legend": a verifiable backstory that connects with the present and validates a subject's inhabitation of a particular community.[1] Rather than begin with a blank slate, officers preferred to base their identities on real people who had died, ideally deceased children whose official records were sparse enough to be added to without fear of contradiction, thus giving a stronger internal coherence to the officers' legends. Proceeding on this basis, "Met officers, without informing the families of the children, and using . . . original birth certificates, built a profile for themselves that would pass for an actual person."[2] Following this model, O'Hagan constructed a profile for Pinn, beginning with his birth certificate, which was obtained with disarming ease from the General Register Office. This initiated a "process of legitimisation" in which further documentation, such as a driver's license, was applied for, and in which the legend was further "grounded" by O'Hagan's rental of a flat in Pinn's name.[3] In collusion with an acquaintance in film production, O'Hagan fashioned a photographic image of an imagined Pinn that could be appended to relevant documentation; those that he could not apply for by normal means were obtained illegally—the traditional shadowy backstreets and under-the-counter transactions updated for the twenty-first century by an excursion into the dark web. As O'Hagan reports, "It wasn't long before I saw Ronnie's face on a driving licence. It took a few weeks to secure a passport. . . . Slowly and digitally, 'Ronnie' began to be a man who had everything: a face, an address, a passport, discount cards."[4]

The idea of identity as a legend, in this sense, speaks to a central concern of this collection: namely, that identity formation is inextricably tied to the act of being identified. Undercover officers' successful inhabitation of a credible identity depends on the ability to realize their chosen identities in a believable manner; they must be recognized as someone who fits into the given group and endorses its prevailing ideology. But for this masquerade to become fully credible it has to be accompanied by a manipulation of the documentation through which individuals are proven to be themselves and no other—having an identity means being identified as such. The traces that lend credence to a given identity are therefore grounded in both the impressionistic sense of who they are and the conglomeration of identificatory presences that prove it to be so.

A legend, in this and its more standard usage, forms a connection in which the present is made explicable through recourse to the past. By engaging with the legend, we understand something intrinsic to a sense of enduring, usually national, identity. But legends are, of course, also fictions that mythologize a past that was itself never meaningful in precisely this way. To form a legend is therefore to merge hard facts with the power found in the representational arts. As O'Hagan puts it, "When I first heard about it, I wondered if the officers involved in this activity were not in fact covert novelists, giving their 'characters' a hinterland that suited the purpose of their present investigations."[5] And it is the power of the imagination that O'Hagan's experiment, written up in an article for the *London Review of Books* in 2015 and as a chapter in his *The Secret Life* (2017), makes plain. For while the traditional forms of paper documentation serve as the originators of "Ronnie," O'Hagan finds that he is unable to sustain a believable identity without embedding his character in the technological manifold through which such an identity is creatively constituted in the contemporary moment. In other words, the identification of Ronnie as a genuine person is wrapped up in a complex skein of traces that include both his formal identification and his online presence within a host of social media and communications technologies. Ronnie thus opens Facebook, Gmail, and Twitter accounts, validating his integration in the world through evidence of his "friends" and "followers" and also through the cultural forms (music, films, books, and so on) that he "likes."

Although O'Hagan's actions here are entirely fictionalizing—"Ronnie," unlike Ronald Pinn, never truly exists—they reflect and depend on widespread social media practices through which users are encouraged to manufacture a particular profile: what O'Hagan refers to as the "invented life."[6] O'Hagan addresses this contemporary push toward fabrication based on partial fact(s) or knowledge by noting that "people nowadays are often obsessed with ancestry, and records that used to take weeks to search are now visible in a matter of minutes, for a fee. At the same time, Facebook and other social media platforms encourage the opposite: the invented life. Writing this story, I moved continually from one way of knowing a person to another, from the real to the fictive, and it seemed a perfectly contemporary way of understanding a life."[7] Of Facebook's 864 million daily users, 67 million are thought to be fake, with many of these not even forming the cover for an actual person but existing only as digital robots formed from complex algorithms that facilitate market research into the preferences of potential consumers.[8] What the invention of Ronnie highlights, then, is not simply the extreme practices of covert operations but the mundane forms in which contemporary self-identity is performed—forms that designate "a change not just in the technological basis of our lives but in the narrative strategies now available to us. You could say that every ambitious person needs a legend to deepen their own."[9]

The creation of Ronnie involves an entangled production that attests to the reciprocal relationship between identification and identity and its embeddedness in a context that is at once technological, bureaucratic, social, and cultural. Like the rest of us, he is known by his driver's license and passport *and* by the fact that his assimilation into technological media is an "invented life." Ronnie's proof of life is multidirectional; the straightforward sense of an identity that is simply verified by a form of identification is complicated by his pursuit of interests, documented in various social media outlets, that follow from his identificatory background(s). For example, Ronnie's immersion in the sites of the dark web, ordering drugs and guns and taking part in extreme discussion forums, sees his potential markers of identity following from the sources of his originating identification.

Introduction

This is no accident. In O'Hagan's view, the web—dark or otherwise—has become the predominant means of self-fashioning. Social media accumulates and shares the traces of an identity with disparate thoughts and whims brought together to form a collected "account" of that person. Yet, when examined within the context of the history of identification, what is noticeable about this process is how similar it looks to a forensic conceptualization of identity. Since the development of fingerprinting in the late nineteenth century, forensic definitions of identity have centered around notions of contact, an approach epitomized by the French criminologist Edmond Locard's influential and oft-quoted dictum that "every contact leaves a trace." The networked technologies of the contemporary world thus increasingly necessitate that legitimacy is formed through the visible manifestation of "contact": online address books are lists of contacts, after all. If we follow the logic of Locard, these traces resist erasure. By the time O'Hagan decides to end Ronnie's brief existence, he has established a tax code and national insurance number, and his business is being solicited by banks—records of which will linger on in the bureaucratic management systems of state and commercial institutions.[10] And although Ronnie's social media accounts can be deleted, they too "leave a shadow on the net" such that "the fake Ronnie is indelible, his 'legend' part of the general ether. He has 'metadata,' the stuff that governments collect, the chaff of being. And he will go on existing in that universe though he never existed on earth."[11]

As Jane Caplan and John Torpey have rightly pointed out, acts of identification have historically been attended by "the uneasy sense that we never fully own or control our identity, that the identity document carries a threat of expropriation at the same time as it claims to represent who we 'are.'"[12] This anxiety, born from an imagining of human selfhood infinite in its richness and entirely resistant to abbreviation, envisages identification(s) as only ever reductive. Yet, what O'Hagan's Ronnie experiment lays bare is the fact that such identifications are never merely routes to a basic unit of data that bears no relation to individual subjectivity. Identities are not reducible to physical data, documentation, or digital information—but neither are identifications. Indeed, the very complexity of identifications—their deep embeddedness in sociocultural practices and histories, imagined realities and fictionalizations—is what renders them productive of such rich and varied conceptualizations of identity.

Placing Identity and Identification in Sociohistorical Perspective

That identifications are socially constructed entities produced through an interaction between participants has been noted by many. Valentin Groebner writes, for example, of identification as "a process that involves more than one person,"[13] and Edward Higgs comments that identification is "not something that resides in the individual but in the social interactions in which they take part."[14] This distinction between the exteriority of identification and the interiority of individual subjectivity is important, but it is also a distinction that is worth troubling. Taking a cue from the work of Kelly Gates and from dominant lines of thought in cultural and critical theory of the past several decades, *The Art of Identification* begins with the premise that such interiority is itself a social construct that can affect and be affected by the similarly constructed nature of identifications. Indeed, Gates accentuates this connection when invoking Stuart Hall, one of the most influential theorists of identity, to claim that identification "is not a fixed and stable object of 'an already accomplished fact,' but a 'production,' which is never complete, always in process, and always constituted within, not outside, representation."[15] Gates is concerned with the obvious ways in which identifications are the product of social encounters—such as the fact that they take place at particular locations, like state borders, and involve an interaction between identifier and identified—but her main focus is on the social dimension to the scientific and technological developments that drive the "production" of identifications. And although she is specifically writing about facial recognition technology, this approach is equally relevant to a whole host of identificatory techniques.

As the chapters in this collection demonstrate, scientific and technological advances in the late twentieth and early twenty-first centuries ostensibly seek to eradicate anonymity. This is particularly evident in the creation and application of digital surveillance, biometric technologies, and DNA typing (see parts 2 and 3). Yet what the volume also reinforces is that this drive toward deanonymization has a prehistory. Valentin Groebner's careful exposition of the many means through which human identification took place in medieval Europe and Natalie Zemon Davis's now iconic excavation of the Martin Guerre story are both testaments to the enduring desire for reliable human individuation.[16] It is a desire that takes on a distinctive shape, however, in the context of a decidedly modern imagining of the relationship between citizen and state. As Caplan and Torpey argue, "It was the epoch of political development inaugurated by the French Revolution's creation of a specifically national citizenship that stimulated the spread of both the resources and the need to subject entire populations to large-scale documentary inventories, and hence the adoption of elaborate systems for tracking and verifying individual identities."[17] Historical work on identification has tended to follow this model, seeing the conditions of Western modernity as generative of the need for human identification and the means with which to make identifications. That said, it is not until the mid- to late nineteenth century that identification as a matter of scientific interest was married to the conditions that allow for its mass bureaucratization and centralization.[18] Gates argues that by the end of the century "standardized documents, archives, and administrative procedures for the management of individual identities itself [had] displaced the more personal and informal forms of trust and recognition characteristic of smaller-scale forms of social organization," while for Patricia E. Chu, it is during the early decades of the twentieth century that "modern states developed unprecedented abilities to identify, track and regulate populations."[19] The historical period and geographic locations covered in *The Art of Identification* are consistent with these models. A congruence between the overarching interest taken in the individual by modern European and North American societies is therefore assumed, and the period covered spans the mid-Victorian era to the present. This decision is not in any way intended to suggest that the history of identification is an exclusively Western affair or to imply that the connection between identification and identity takes place only in modern nation-states.[20] It is, rather, a pragmatic choice intended to foster as much coherence as possible between the various disciplines that are showcased in the ensuing chapters.

Genres, Body Types, and Surveillant Technologies

As we have already implied, identification has recently been a topic of significant academic study. Most of this research has been in the form of social histories that have employed sociological thinking within a context of the history of science to illuminate a number of identificatory methods and their uses. Prominent examples include John Torpey's *The Invention of the Passport: Surveillance, Citizenship and the State* (2000); Jane Caplan and John Torpey's *Documenting Individual Identity: The Development of State Practices in the*

Modern World (2001); Simon A. Cole's *Suspect Identities: A History of Fingerprinting and Criminal Identification* (2002); Craig Robertson's *The Passport in America: The History of a Document* (2010); Edward Higgs's *Identifying the English: A History of Personal Identification 1500 to the Present* (2011); Ilsen About, James Brown, and Gayle Lonergan's *Identification and Registration Practices in Transnational Perspective: People, Papers and Practices* (2013); and Keith Breckenridge and Simon Szreter's *Registration and Recognition: Documenting the Person in World History* (2012).[21] Unsurprisingly, given its disciplinary base, what motivates the majority of this work is an interest less in the individual that is identified and more in the society that identifies. Breckenridge and Szreter, for example, write that "the primary task of historical scholarship is to demonstrate the many ways in which legal forms of identity documentation reflect the power and will to knowledge of the state."[22]

What can be usefully added to this valuable body of work is a precise account of how identification is entangled in a wider culture of identity formation. This connection has often been referred to in passing; Cole, for instance, writes that "identification practices, clearly, produce identifications *and* identities," while Caplan and Torpey comment that "the evolution of procedures for individual identification was . . . tied up with the more recent historical emergence of modern concepts of individuality and subjectivity" but has not yet formed the subject of sustained examination.[23] *The Art of Identification* expands the broadly historical approach of existing scholarship to ask questions about how identification has been figured not only by historically situated bureaucracies and sciences but also within the contexts of literature, visual art, and other cultural forms.

Along with the historical interest in identification, the past two decades have witnessed an explosion of interest in literature and the forensic sciences.[24] From scholarship on detective fiction to the *CSI* television franchise to the art of DNA profiling, literature and science scholars have established and catered to an audience craving books about the intersections between forensics and a multitude of media.[25] Ronald R. Thomas's *Detective Fiction and the Rise of Forensic Science*, a foundational work in this field, is deeply interested in how certain forensic techniques (including fingerprinting) conceptualize and imagine the individual in unique ways, with these techniques providing not just the method by which late nineteenth-century and early twentieth-century fictional detectives solve cases but also the means by which authors and publics construct their representations of the detective and criminal. For Thomas, the late nineteenth century thus witnessed "the systematic transformation of the notion of the individual citizen's essential reality from something we call 'character' to something we came to call 'identity'" with the various modes of reading such identity then becoming an essential constituent of modern experience.[26]

Because identifications are typically premised on reading a *sign* of identity, the alluring prospect of considering identificatory technologies to be operating as systems of signs is also opened. Jane Caplan gestures toward this approach when pointing out how three of the clearest modes of modern identification—the name, portrait, and fingerprint—"correspond, by a logic that is surely not accidental, to C. S. Peirce's 'second trichotomy of signs,' namely, his differentiation among symbol, icon, and index according to the sign's own character."[27] Peirce's influential "trichotomy" distinguishes between a sign that forms an arbitrary connection to its object and functions only because of habitual usage (symbol), a sign that resembles its referent in some way (icon), and a sign that is linked to its object through a dynamic connection (index). When mapped onto a temporal context, the trichotomy offers a model for reading the history of identificatory practices in which the ever-greater demand to know the individual has been met by techniques that have not relied on arbitrary information (such as a name) but first on technologies of resemblance such as the photograph and ultimately on the use of indexical approaches such as anthropometry, fingerprinting, and DNA. For Kata Gellen, it is this latter move that represents something decidedly modern—namely, the beginning of "a view of individual identity that is determined not by personality or interiority but by the physical traces a person leaves behind" and illustrates "how identity is conceived and established in modern society—as something cold, objective, and impersonal."[28]

The trichotomy of signs, especially the dichotomy between symbol and icon on the one hand and index on the other, similarly resonates with Higgs's perceptive account of how identification has, traditionally, deployed a different class of techniques depending on the category of individual in question. According to Higgs, "When identifying themselves as juridical persons, who are able to claim the right to acquire and alienate property, individuals have historically used possessions they could produce, acts they could perform, or signs they could utter (e.g., the use of seals, signatures or of PIN numbers in electronic transactions)."[29] Criminals, by contrast, have tended to be identified by their physical beings, whether that be through the body's treatment as a canvas on which signs of criminality could be indelibly printed (branding, disfigurement, tattooing) or its conceptualization as an archive of its own permanent traces (fingerprints, DNA profiles).

These broad characterizations of identification are useful but incomplete. So, in Higgs's own estimation, the historical separation between citizen and criminal has been weakened by recent developments that have seen the body—conceptualized as machine-readable biometric data—become a much more prominent part of the identification of employees, customers, and passengers.[30] And while on the face of it these developments would appear to corroborate the notion of a narrative line that moves from symbol, through icon, to index, this idea breaks down in the context of real-world identifications, a point made clear in the example of "Ronnie Pinn," whose validating, interlocking set of credentials were acquired on the basis of an originating document that is notoriously easy to forge—the "symbolic" birth certificate.[31] This sense that the act of identification is premised on a kind of provenance in which one moment of identification is related back to an original moment of authentication is, again, linked to the idea of a system of signs, in this instance the notion that identification relies on a diachronic matching. But at the same time, identification can *only* work when housed within a synchronic system that deems each validating and original identification to be essentially different from every other. Identifications thus work on the dual basis that the identificatory data in question are demonstrably identical to those contained in another record *and* that the record itself is clearly distinguishable from all other identificatory records. In this latter context, the origin of the validating record is less important than its integration within a system of difference—as is suggested by Cole's evocative description of fingerprints as "empty signifiers" that say nothing other than marking a differentiation from other such patterns.[32] For scholars and students of the humanities (especially literature), the resonance here with a Saussurean conceptualization of language as a system of arbitrary signs is obvious. To their way of thinking, identifications take place within a system of writing that renders people "legible" in multiple ways.[33]

Many of the chapters in our collection speak to the coherencies and tensions that exist between the legibility afforded by various practices of identification and the numerous other cultural modes that render individuals legible. Indeed, the chapters in part 1 have been grouped under the title "Genres of Identification" to reflect how they, in particular, trace the often surprising connections between diverse cultural forms, including genre fiction, and identification. But if genres are to be taken as forms that create normative expectations of content—in other words, that impose limitations on what kinds of identities emerge within them—then it is not just the obvious examples of noir thriller or whodunit that can be classed as genres. Rather, there is a whole host of genres that make and remake identity, with modes of identification being one of them.

Take Matt Houlbrook's opening chapter, for instance, which examines a fleeting moment from the end of the Great War to the late 1920s when identification is posited as a "problem in a plausible world." Contemporary scholarship is used to thinking about such problems as a challenge for state bureaucracies and national security, but Houlbrook's focus is much more on how it infiltrates networks of commerce and leisure and blurs the boundaries between state, society, and culture. Two particular genres of identification are Houlbrook's focal points: (1) the existence of newspaper competitions that assessed the "character" of submitted photographs and the growing number of commercial enterprises, such as the Pelman Institute and ABC Correspondence Schools, that offered to train students in the art of such "identifications" and (2) the conflicting programs in personal transformation that were offered by similar organizations and that promised an education in the tricks required to attain the confidence of those one met. For Houlbrook, the 1920s were thus distinguished by the belief in two incompatible ideas: that character could be read from appearance, while personality could be refashioned. And it is this tension that turns the analysis proffered into more than a mapping of the traffic from criminological and anthropological modes of identification to high, middlebrow, and popular culture. Rather, those latter modes, in disrupting the "sequence and object" of processes of identification, trouble the capacity to locate individuals during the interwar period and speak to a specific constellation of cultural circumstances.

The expectation that genre fiction will provide repeated representations of certain character types is perhaps nowhere more prevalent than in noir, where the world-weary private eye and femme fatale are ubiquitous. Yet in James Purdon's chapter, "Identity Noir," the move from the hard-boiled novel of the 1920s and 1930s to the noir thriller of the 1940s does not fix identities but paradoxically charts an emerging identity crisis wrought by the proliferating modes through which a public identity could be secured. Noir's dominant position as *the* form of American crime writing therefore arises just as new institutions and bureaucratic protocols for registering and recording identity are emerging. Yet, while the bureaucratic and technological means of making an identification helped to drive the plot and establish the gritty verisimilitude of novels by Dashiell Hammett and Raymond Chandler, it is the existential crisis they prompt that characterizes the less well-known but no less significant noir of writers such as Cornell Woolrich and Kenneth Fearing. In Fearing's *The Big Clock* (1946), it is therefore the protagonist's confrontation with his identifiable self, brought about through an ingenious plot of self-pursuit, that leaves him questioning the authenticity of his own selfhood. As Purdon's analysis demonstrates, late noir as a genre is predicated on the paradoxical condition of narrating the jarring encounter with nonnarrative identity.

In Andrew Mangham's chapter, by contrast, it is the very narrative structure of forensic science that is the concern. Mangham describes the nascent forensic science of the mid-nineteenth century as a genre concerned with turning trace evidence into narratives of past events. As such, the medical jurisprudence of the time shares distinct similarities with the period's literature, especially that of Charles Dickens. Beginning his analysis with the discovery, on 9 October 1857, of a bag filled with dismembered remains and the attempted reconstruction of its origins by Scotland Yard and the famed medic Alfred Swaine Taylor, Mangham then focuses on the many ways in which Dickens's plot lines and character portraits revolve around a similarly forensic reconstruction. Mangham provides a further interconnection by bringing fresh scholarly interest to the composite photography of Francis Galton. The chapter problematizes the received logic positing that Galton was pursuing a project that he conceived in simplistically objective terms. Rather, for Mangham, Galton's photographs and his thinking about them evince a thoughtful, self-reflexive approach to method that brings Dickens's own doubts about the frailties of realism into sharp relief. Rather than inserting details about medical jurisprudence and its reading of identifications to promote a realistic credence, Dickens's work is more perceptively thought of as mirroring the emphatically fictional implications brought about when acts of forensic identification are pursued.

Prose fiction has, of course, often found more explicit means of addressing identification in the context of crime and policing. Detective fiction, as it developed during its golden age of the interwar years, presented the crime itself as a puzzle that required the logical calculations of the fictional detective—and the reader—to solve. But in the 1930s a new style of detective story emerged, as Victoria Stewart shows in her chapter on the writing of Francis Iles, which displaced the puzzle element with the question of authorship (Iles was a pseudonym) and shifted the focal point of the narrative from the detective to the criminal. In subtly deviating from the generic conventions of the detective novel and venturing more into explicatory nonfiction and the thriller, Iles turned the task of identification to the complexities of motive. Drawing on the popularization of psychoanalytical models of thought, Iles's novels therefore lie in the space between a criminological and a medical understanding of crime and anticipate the writings of post–World War II writers such as Patricia Highsmith. The problem of identification is retained in this reshaping of genre but in altered terms. Rather than identification of the culprit, the issue is the morally thorny one of how far identification *with* the criminal should take place.

Stewart's interest in the criminal as both individual and category makes her chapter an appropriate bridge into part 2 of the collection. Taking the representations afforded by and made of a variety of sciences and technologies of identification, this section builds on Jane Caplan's assertion that "in virtually any systematics of identification, everyone is not only 'himself' but also potentially the embodiment of a type."[34] In an abstract sense this is perhaps unsurprising: as much as identifications differentiate us from all others they also tend to place us in a series that renders identities through the same limited set of data points. But the history of identification is also littered with examples of more sinister projects of classification. For example, the early history of fingerprinting as it was developed in British India and especially as it was presented by Francis Galton in the 1890s often betrayed problematic assumptions about racial difference. Galton pursued the paradoxical and equally offensive goals of creating a system that would allay "the difficulty felt by most Europeans in accurately distinguishing the features of men of the darker races" and of suggesting that fingerprints provided a key to establishing the intrinsic difference between entire races.[35] In more recent history, DNA profiling's reliance on population genetics and the perceived potential of genetic science has at times aroused a similarly troubling rhetoric.[36]

Individual bodies are also fundamentally shaped by elements of the lived environment that are shared in uneven ways with other members of the population. Rebecca Gowland and Tim Thompson turn to this subject in a chapter that could easily be called the "Social Life of the Skeleton," as they detail the myriad ways that skeletons and skeletal remains are complicated by sociocultural factors both in life and after death. For them, skeletal identification practices are firmly situated in a social milieu complicated by issues of diet, environment, and even maternal health and by the language, training, and technologies available to researchers working with the remains. By training, Gowland and Thompson are a bioarchaeologist and a forensic anthropologist, respectively. While both scientists acknowledge that bioarcheaology is a more socially situated discipline than its forensic cousin, they argue that seemingly objective questions about human remains (including, for example, age and date of death) are not straightforward. As feminist science studies theorists have argued for the past few decades, objectivity is an ideal and a myth that could and should be set aside in favor of more situated knowledges. What is fascinating about this chapter are the ways that Gowland and Thompson make this case from *within* their respective scientific disciplines, arguing for situated sociocultural thinking on the part of scientists addressing skeletal identification. They urge bioarchaeologists—and perhaps especially forensic anthropologists—to embrace the messiness of partial complex conclusions instead of seeing such narratives as weakening the field's disciplinary authority.

If Gowland and Thompson bring sociocultural issues to bear on bioarchaeology and forensic anthropology, Patricia E. Chu's chapter raises questions about reading ethnoracial novels in a postgenomic era. Her focus is on Ruth Ozeki's novel *My Year of Meats*, but she uses this narrative and her analysis of it as a springboard to discuss contemporary intersections between genetics, race, corporations, and the environment. In particular, she argues that our current postgenomic bioculture may be particularly challenging for authors of ethnoracial novels who have questioned and will continue to question essentialist explanations of identity. In a world in which just about everything—including our genetic material—has been annexed by corporations, *how* do we identify ourselves? When DNA becomes our life's code, what are the consequences of allowing biosubjectivity to (re)define us, even if there is no gene for race? Chu's reading of *My Year of Meats* is an open-ended and somewhat hopeful demonstration that the ethnoracial novel has multiple avenues by which to navigate a postgenomic bioculture and that active engagement with genomic data can result in continual reinterpretation of our identificatory codes.

Rounding off this part of the collection is Jonathan Finn's chapter, which traces how the very public arrest and trial of Colonel Russell Williams was shaped by the predominantly visual means through which his criminality was identified. Directly addressing the photograph as identificatory evidence, Finn uses the Williams case as a context in which to pose important questions about "ubiquitous photography" and "everyday aesthetics." By tracing the often explicit photographic evidence collected by Williams, deployed by the courts, and marshaled by the media, Finn argues that a particular visual culture informs this Canadian crime spree of sexual assault. Looking to W. J. T. Mitchell's "visual turn" as a backdrop for his case, Finn demonstrates the increasing importance of the visual for criminalistics in an age of multimodal, often social, media. Moreover, he links this specific upsurge in crime and media to the unfortunate commonplace of gendered violence. The photograph, then, remains a particularly potent symbol and tool of identification, even as it has taken on new meaning and life in the twenty-first century.

Williams's photographs of himself laid his criminality bare for everyone to see. But what happens when what the photograph proves is measured by digital device rather than the human eye? In Liv Hausken's essay on the biometric passport, she questions the addition of facial recognition technology (FRT) to passports around the globe. Providing a close reading of *Doc 9303: Machine Readable Travel Documents*, a report issued by the International Civil Aviation Organization (ICAO), Hausken argues that this document builds a case for FRT based on social acceptance and practical efficiency: FRT is similar enough to photographic identification requirements and allows for both the identification of individuals as distinct agents and the comparison of individuals to information in databases, such as watch lists. By reading *Doc 9303* literally forward and backward, Hausken's analysis offers insight into the elision of photography and FRT, all the while considering the dubious means through which FRT was introduced and ostensibly accepted without debate. Indeed, many passports have become multimodal biometric documents, and as Hausken argues, the groundwork for public acceptance has already been laid. Her chapter is—for, we hope, obvious reasons—an appropriate opening to part 3 of our collection: "Surveillant Technologies."

The idea, referred to earlier, of making identifications within a ceaseless present of synchronic differentiation accentuates the mutual imbrication of contemporary identificatory practices and the wider field of surveillance. Much recent work on identification has demonstrated that this affinity has a prehistory. John Torpey's *The Invention of the Passport: Surveillance, Citizenship and the State* (2000); Valentin Groebner's *Who Are You? Identification, Deception and Surveillance in Early Modern Europe* (2007); and Jonathan

Finn's *Capturing the Criminal Image: From Mug Shot to Surveillance Society* (2009) even contain *Surveillance* in their titles, while works more clearly designated as part of a burgeoning "surveillance studies" (e.g., David Lyon's *Surveillance Society: Monitoring Everyday Life* [2001]) have clearly been concerned with identification.[37] Engaging with the logic of surveillance as a means to interpret cultural artifacts has been done before—as long ago as John Bender's *Imagining the Penitentiary: Fiction and the Architecture of Mind in Eighteenth-Century England* (1987) and as recently as David Rosen and Aaron Santesso's *The Watchman in Pieces: Surveillance, Literature, and Liberal Personhood* (2013)—but never with the intention of interrogating how the specific feature of identification is embroiled in identity's inhabitation of surveillance's imaginary.[38]

The boom in recent scholarship on surveillance as a social, political, and cultural phenomenon has been stimulated by the rapid expansion, since the beginning of the twenty-first century, of a wider body of state-mandated surveillant technologies and practices such as CCTV recording, security screening at transport hubs, and the collection of detailed metadata concerning telephone and online communications. Increasingly sophisticated and increasingly visible, the methods by which the state monitors various sources of surveillant information on individuals and groups have raised justified concern among scholars, as among the general public, regarding the interaction of such forms of surveillance with human-rights law and long-standing conventions of civil liberty. Yet the top-down story of the "surveillance society," in which states strive for ever-tighter control of populations, constitutes only one thread in the mutually implicated histories of identification and surveillance. The desire to identify and monitor others is, in fact, more widespread and complex than can adequately be accounted for by paradigms shaped by the perceived disciplinary control of official agencies and central governments.

Such complications highlight what Rex Ferguson, in a chapter on the Jason Bourne series of Hollywood thrillers, calls the "messy conditions" faced by contemporary subjects whose "desire to be free" is conjoined with a no less insistent "desire to be named." Resisting the common association of identification with authoritarian control, Ferguson shows how the Bourne films problematize the standard theoretical division of identity between, on the one hand, the memories and experiences of a continuous embodied self and, on the other hand, the alienated representation of the body and its historical traces in the archives of the state. Unwilling to accept his officially imposed identity yet equally unable to discard it in favor of a more authentic self, Bourne becomes representative of the modern individual, caught impossibly between multiple, often conflicting, forms of identity.

As the Bourne films demonstrate, the desire to be known and seen as a unique individual exists in tension with the everyday reality of social arrangements that encourage and at times necessitate the formation of group identities, whether by imposition (as in bureaucratic categorizations of racialized or gendered difference) or by association (as in, for instance, the sometimes contentious negotiation of collective identity/ies within the LGBTQ+ community). The recognition of that tension—between the principle that every individual has a right to an identity and the unavoidable fact that all identities are fundamentally social formations—has been a key theme of utopian and dystopian fiction since at least H. G. Wells, in part because utopian visions seem always to imply some weakening of individual identity as a prerequisite for social harmony. Dorothy Butchard's chapter focuses on two recent dystopias—J. G. Ballard's *Super-Cannes* (2000) and Nicola Barker's *H(A)PPY* (2017)—to show how fiction can be used to dissect the technological-supremacist rhetoric of the "intelligent city." Discussing these novels, whose inhabitants willingly consent to round-the-clock monitoring by systems of identification, surveillance, and quantification as the price of living in stable, utopian societies, Butchard explores their respective critiques of "shareveillance" and highlights their authors' attentiveness to the "wayward body" as the location of violent or subversive affects that might pose a challenge to social cohesion.

Whereas Butchard emphasizes the body as a site of affective resistance to authoritarian impositions of identity, Robert Lederer's chapter on Jennifer Egan's novel *A Visit from the Goon Squad* (2010) finds another possible escape route in material archives. If contemporary accounts of identity conjure visions of a world populated by data doubles, in which our identity resides not in our bodies but in an aggregate of data points distributed between countless electronic databases, Lederer finds in Egan's novel both a formal engagement with the random-access structure of such databases and a countervailing interest in how personal collections of material objects might constitute another sort of distributed identity, an archive whose material form renders it opaque to uninitiated readers.[39] Setting the digital database, structured around the principles of quick access and unambiguous legibility, against the mysterious and idiosyncratic assemblage of collected objects that require interpretation,

Lederer shows how Egan's characters seek to carve out "a space of personal opacity" from the demands of technological identification systems.

The chapters in our collection reflect scholarly work from a range of disciplines and attend to a rich variety of artifacts taken from the representative arts, the historical record, and contemporary policy. These artifacts at times affirm and at times problematize but always demonstrate the mutual imbrication between processes of identification and the cultural formation of identity. What we hope to prove, in the pages that follow, is not merely that processes of identification are a significant context in which representations of identity take place. Our intention, rather, is to reflect how identification's myriad means of creating interpretable traces of identity are imaginative acts of representation in themselves—acts that cohere with and grate against similarly imaginative acts of identity formation. That there is such a thing as an art *about* identification is one thing. The more pressing claim of this collection is that there is an art *of* identification.

Notes

1. "A 'legend' is KGB jargon. It means a false biography, a cover story." John le Carré, in Matthew J. Bruccoli and Judith S. Baughman, *Conversations with John le Carré* (Jackson: University Press of Mississippi, 2004), 68.
2. Andrew O'Hagan, *The Secret Life: Three True Stories* (London: Faber and Faber, 2017), 102.
3. Ibid., 109.
4. Ibid., 118.
5. Ibid., 102.
6. Ibid., 110.
7. Ibid.
8. Ibid., 114.
9. Ibid., 116.
10. Ibid., 129.
11. Ibid., 127.
12. Jane Caplan and John Torpey, eds., *Documenting Individual Identity: The Development of State Practices in the Modern World* (Princeton: Princeton University Press, 2001), 8.
13. Valentin Groebner, *Who Are You? Identification, Deception, and Surveillance in Early Modern Europe* (New York: Zone Books, 2007), 27.
14. Edward Higgs, *Identifying the English: A History of Personal Identification 1500 to the Present* (London: Continuum, 2011), 13.
15. Kelly A. Gates, *Our Biometric Future: Facial Recognition Technology and the Culture of Surveillance* (New York: New York University Press, 2011), 16.
16. Groebner, *Who Are You?*; Natalie Zemon Davis, *The Return of Martin Guerre* (Cambridge, MA: Harvard University Press, 1983).
17. Caplan and Torpey, *Documenting Individual Identity*, 7.
18. Higgs, *Identifying the English*, 141.
19. Patricia E. Chu, *Race, Nationalism and the State in British and American Modernism* (Cambridge: Cambridge University Press, 2006), 2; Gates, *Our Biometric Future*, 13.

20. For an international perspective, see Ilsen About, James Brown, and Gayle Lonergan, eds., *Identification and Registration Practices in Transnational Perspective: People, Papers and Practices* (London: Palgrave Macmillan, 2013); and Keith Breckenridge and Simon Szreter, eds., *Registration and Recognition: Documenting the Person in World History* (Oxford: Oxford University Press, 2012).

21. John Torpey, *The Invention of the Passport: Surveillance, Citizenship and the State* (Cambridge: Cambridge University Press, 2000); Caplan and Torpey, *Documenting Individual Identity*; Simon A. Cole, *Suspect Identities: A History of Fingerprinting and Criminal Identification* (Cambridge, MA: Harvard University Press, 2001); Craig Robertson, *The Passport in America: The History of a Document* (Oxford: Oxford University Press, 2010); Higgs, *Identifying the English*; Ilsen About, James Brown, and Gayle Lonergan, eds., *Identification and Registration Practices in Transnational Perspective: People, Papers and Practices* (London: Palgrave Macmillan, 2013); Breckenridge and Szreter, *Registration and Recognition*.

22. Breckenridge and Szreter, *Registration and Recognition*, 21.

23. Caplan and Torpey, *Documenting Individual Identity*, 2; Simon A. Cole, "De-Neutralizing Identification: S. and Marper V. United Kingdom, Biometric Databases, Uniqueness, Privacy and Human Rights," in *Identification and Registration Practices in Transnational Perspective: People, Papers and Practices*, ed. Ilsen About, James Brown, and Gayle Lonergan (London: Palgrave Macmillan, 2013), 77.

24. In what is less clearly a work about literature and forensics—yet built on a foundational appreciation of detail that could be said to form a prehistory of the science— Kamilla Elliott's *Portraiture and British Gothic Fiction: The Rise of Picture Identification 1764–1835* is similarly concerned with genre, in this case that of Gothic novels and portrait painting. Moving beyond a simplistic noting of the moments when identifications are made in Gothic texts and how those moments are facilitated by portraits, Elliott's concern is with the functioning of "mimetic resemblance" across both forms. Portraiture and realist literary form thus generate certain kinds of identity as opposed to merely documenting them. Yet even within this frame, Elliott's stated concern is less "with the formation of the *subjective* modern self than with cultural practices and mythologies of *social* identification" (25).

25. See, for example, Melissa M. Littlefield, *The Lying Brain: Lie Detection in Science and Science Fiction* (Ann Arbor: University of Michigan Press, 2011); Kata Gellen, "Indexing Identity: Fritz Lang's M," *Modernism/Modernity* 22, no. 3 (2015); Andrew Mangham, *Dickens's Forensic Realism: Truth, Bodies, Evidence* (Columbus: The Ohio State University Press, 2017); and Victoria Stewart, *Crime Writing in Interwar Britain: Fact and Fiction in the Golden Age* (Cambridge: Cambridge University Press, 2017).

26. Ronald R. Thomas, *Detective Fiction and the Rise of Forensic Science* (Cambridge: Cambridge University Press, 1999), 11.

27. Jane Caplan, "'This or That Particular Person': Protocols of Identification in Nineteenth-Century Europe," in *Documenting Individual Identity: The Development of State Practices in the Modern World*, ed. Jane Caplan and John Torpey (Princeton: Princeton University Press, 2001), 52.

28. Gellen, "Indexing Identity," 426.

29. Higgs, *Identifying the English*, 13.

30. See chapters 8 and 9 in this book.

31. James Rule et al., "Identification and Mass Surveillance in the United States," *Social Problems* 31, no. 2 (1983): 227; Roger Clarke, "The Digital Persona and Its Application to Data Surveillance," *The Information Society* 10 (1994): 87.

32. Cole, "De-Neutralizing Identification," 84.

33. Caplan and Torpey borrow this phrase—"legible people"—from James Scott. Caplan and Torpey, *Documenting Individual Identity*, 1, 6.

34. Caplan, "This or That Particular Person," 51.
35. Francis Galton, "Identification," *The Times*, 7 July 1893; Cole, *Suspect Identities*, 36.
36. Jay D. Aronson, *Genetic Witness: Science, Law and Controversy in the Making of DNA Profiling* (Piscataway: Rutgers University Press, 2007), 207.
37. David Lyon, *Surveillance Society: Monitoring Everyday Life* (Buckingham: Open University Press, 2001).
38. John Bender, *Imagining the Penitentiary: Fiction and the Architecture of Mind in Eighteenth-Century England* (Chicago: University of Chicago Press, 1987); David Rosen and Aaron Santesso, *The Watchman in Pieces: Surveillance, Literature, and Liberal Personhood* (New Haven: Yale University Press, 2013).
39. Kevin D. Haggerty and Richard V. Ericson, "The Surveillant Assemblage," *British Journal of Sociology* 51, no. 4 (2000).

References

About, Ilsen, James Brown, and Gayle Lonergan, eds. *Identification and Registration Practices in Transnational Perspective: People, Papers and Practices.* London: Palgrave Macmillan, 2013.

Aronson, Jay D. *Genetic Witness: Science, Law and Controversy in the Making of DNA Profiling.* Piscataway: Rutgers University Press, 2007.

Bender, John. *Imagining the Penitentiary: Fiction and the Architecture of Mind in Eighteenth-Century England.* Chicago: University of Chicago Press, 1987.

Breckenridge, Keith, and Simon Szreter, eds. *Registration and Recognition: Documenting the Person in World History.* Oxford: Oxford University Press, 2012.

Caplan, Jane. "'This or That Particular Person': Protocols of Identification in Nineteenth-Century Europe." In *Documenting Individual Identity: The Development of State Practices in the Modern World*, edited by Jane Caplan and John Torpey, 49–66. Princeton: Princeton University Press, 2001.

Caplan, Jane, and John Torpey, eds. *Documenting Individual Identity: The Development of State Practices in the Modern World.* Princeton: Princeton University Press, 2001.

Chu, Patricia E. *Race, Nationalism and the State in British and American Modernism.* Cambridge: Cambridge University Press, 2006.

Clarke, Roger. "The Digital Persona and Its Application to Data Surveillance." *The Information Society* 10 (1994): 77–92.

Cole, Simon A. "De Neutralizing Identification: S. and Marper V. United Kingdom, Biometric Databases, Uniqueness, Privacy and Human Rights." In *Identification and Registration Practices in Transnational Perspective: People, Papers and Practices*, edited by Ilsen About, James Brown, and Gayle Lonergan, 77–97. London: Palgrave Macmillan, 2013.

———. *Suspect Identities: A History of Fingerprinting and Criminal Identification.* Cambridge, MA: Harvard University Press, 2001.

Davis, Natalie Zemon. *The Return of Martin Guerre.* Cambridge, MA: Harvard University Press, 1983.

Elliott, Kamilla. *Portraiture and British Gothic Fiction: The Rise of Picture Identification 1764–1835.* Baltimore: Johns Hopkins University Press, 2012.

Galton, Francis. "Identification." *The Times*, 7 July 1893.

Gates, Kelly A. *Our Biometric Future: Facial Recognition Technology and the Culture of Surveillance.* New York: New York University Press, 2011.

Gellen, Kata. "Indexing Identity: Fritz Lang's *M*." *Modernism/Modernity* 22, no. 3 (2015): 425–48.

Groebner, Valentin. *Who Are You? Identification, Deception, and Surveillance in Early Modern Europe*. New York: Zone Books, 2007.

Haggerty, Kevin D., and Richard V. Ericson. "The Surveillant Assemblage." *British Journal of Sociology* 51, no. 4 (2000): 605–22.

Higgs, Edward. *Identifying the English: A History of Personal Identification 1500 to the Present*. London: Continuum, 2011.

Littlefield, Melissa M. *The Lying Brain: Lie Detection in Science and Science Fiction*. Ann Arbor: University of Michigan Press, 2011.

Lyon, David. *Surveillance Society: Monitoring Everyday Life*. Buckingham: Open University Press, 2001.

Mangham, Andrew. *Dickens's Forensic Realism: Truth, Bodies, Evidence*. Columbus: The Ohio State University Press, 2017.

O'Hagan, Andrew. *The Secret Life: Three True Stories*. London: Faber and Faber, 2017.

Robertson, Craig. *The Passport in America: The History of a Document*. Oxford: Oxford University Press, 2010.

Rosen, David, and Aaron Santesso. *The Watchman in Pieces: Surveillance, Literature, and Liberal Personhood*. New Haven: Yale University Press, 2013.

Rule, James, Douglas McAdam, Linda Stearns, and David Uglow. "Identification and Mass Surveillance in the United States." *Social Problems* 31, no. 2 (1983): 222–34.

Stewart, Victoria. *Crime Writing in Interwar Britain: Fact and Fiction in the Golden Age*. Cambridge: Cambridge University Press, 2017.

Thomas, Ronald R. *Detective Fiction and the Rise of Forensic Science*. Cambridge: Cambridge University Press, 1999.

Torpey, John. *The Invention of the Passport: Surveillance, Citizenship and the State*. Cambridge: Cambridge University Press, 2000.

Part 1
Genres of Identification

1.
Charming Faces and the Problem of Identification

Matt Houlbrook

By 1925 Stella Pierres must have been accustomed to winning. She was a noted Irish beauty, and success in the *Daily Sketch*'s search for Britain's Venus de Milo had brought her fame and a chance to perform for the crowds thronging the British Empire Exhibition's Palace of Beauty. Yet Pierres was pleased enough with the outcome of the Charming Faces competition, used to publicize the new weekly story paper *Joy*, that she pasted the ornate certificate she received into her scrapbook. Next to a photograph and signature of film star Ivor Novello, the copperplate text read as follows: "I have much pleasure in stating that, judging by the photograph submitted, in my opinion Stella Pierres belongs to the Magnetic Type." Her name and "Magnetic" had been added by hand.[1]

Novello's assessment was not made on aesthetic terms, and it did not measure Pierres's anthropometric similarity to a physical ideal. Instead it located her within a system of character types. Running over several months, *Joy*'s publicity elaborated the "six types of beauty" identified by this "keen student ... [who] has learned to analyze and classify charm at a glance."[2]

> GRECIAN (Noble and Inspiring); ATHLETIC (Stimulating and Breezy); QUAINT (Impish and Unusual); MERRY (Irresponsible and Bewitching); MAGNETIC (Mysterious and Intriguing); INTELLECTUAL (Showing Character and Brains).[3]

If sent a photograph, Novello would "tell you to which of these types you belong." That brought its own rewards, but there were more tangible prizes: an "attractive certificate," "300 sets of Schappe Silk Lingerie," and "six cash prizes of £25" for those nearest the "ideal of each type."[4] The potential payoff of entering exceeded material gain, however. Classification would bring self-knowledge, which in turn contained alluring possibilities for excitement and social success:

> To know your special charm is to double it. To double your charm is to see and know fresh possibilities for chances and adventures in life.
> Ivor Novello has the key.
> Why not use it?[5]

This entreaty was hugely successful: by the end of May, *Joy* had reprinted photographs of dozens of entrants, filling pages with lists of winners and full-page portraits of the overall victors. Pierres, sadly, was not among them.[6]

Far from unusual, Charming Faces was a familiar media stunt. It was animated by the challenges of launching a title at a ferociously competitive moment in the newspaper and magazine industry and was rooted in burgeoning cultures of celebrity, cinemagoing, and beauty. Like similar stunts in the United States, the competition gestured toward a wider obsession with identifying character from nuances of expression, details of demeanor, and the look of a face. Whereas the language of types was unusual, the connection between appearance and character was commonplace.[7] In magazines like *Joy* and newspapers like the *Sunday News*, invitations to submit oneself to judgment or assess the character of strangers went hand in hand. Whether scrutiny turned outward or inward, such competitions ensured the prominence of dilemmas of identification after the Great War: who am I—and who, in turn, are they?

Why did these questions become ubiquitous in 1920s Britain? Taking Charming Faces as a prompt, this essay traces how similar dilemmas played out across boundaries between state, society, and culture and through ostensibly discrete cultural forms. I follow the proliferation of portraits such as those sent to *Joy*. The use of photographs in character analysis had been a feature of newspapers and periodicals since at least the 1890s. The 1920s, however, witnessed a significant broadening of this visual form's presence in commercial print culture. The historian Michael Saler characterizes this broadening as the "spectacularization of everyday life," a process through which the "sheer onslaught of images, representations, and symbols" made the facial portrait ubiquitous and inescapable.[8] Close-cropped photographs, often annotated to show the correlation between appearance and character, featured in competitions to gauge character, identify beauty, and find new screen stars; these competitions focused critical appraisal of actors' skills. Such photographs also figured in the didactic literature of self-improvement and in ads selling correspondence courses in effective speaking to the anxious or aspirational. Charming Faces was echoed in materials produced by educational organizations such as the Pelman Institute and ABC Correspondence Schools. Recognizing the common motifs in which they traded allows us to see unexpected resonances between the feminized realm of mass culture and commercial self-improvement.

In exploring the reach of these dilemmas, I also want to make a bigger argument about the practice of cultural history. It was no accident that facial portraits illustrating competitions and didactic literature were reminiscent of the passport photo and police mug shot. Their contemporaneous visibility reflected the growing use of mechanical and bureaucratic modes of identification by state institutions, new technologies enabling mass reproduction of images, and the dramatic expansion of newspaper and magazine publishing. Although contemporaries were well aware that analysis of expression and physiognomy had much longer—and often discrete—histories, the particular genealogy of the correlation between character and appearance in the 1920s encompassed the sciences of identification consolidated in the late nineteenth century, which reshaped fictional writing and state institutions.[9] Scrutiny of physiognomy and expression was primarily associated with the work of criminal investigation and anthropological study—forms of expertise that sought to categorize and control. In its look and logic, Charming Faces gestured toward adjacent sciences including phrenology, Sir Charles Bell's and Charles Darwin's theorization of emotional expression in humans and animals, Alphonse Bertillon's anthropometry and card indexing, Francis Galton's "composite portraiture" of criminals and Jewish persons, and the Criminal Record Office's accumulation of mug shots.[10] Through the paraphernalia of empirical research, scientific rules, and technical apparatus, these sciences claimed authority to isolate the "truth" of character.

Just as phrenology became a Victorian parlor game, so it is tempting to treat Charming Faces as a popular translation of the sciences of identification; in eliding or ignoring traditional distinctions between physiognomy and expression, the task of assessing facial features became an entertaining puzzle to attract readers and sell magazines. It is true that competitions and self-styled "character experts" pastiched Bertillon's anthropometric measurements. Noting these similarities gets us only so far, however. Scholars across disciplines have observed the porous boundaries between popular and middlebrow culture and between journalism, criminology, law, and the state after this period; the exchange between "true" crime and crime "fiction" is now commonplace. Rather than just describe this process, however, we need to explain why the business and pleasure of identification assumed this particular form and understand the significance of these congruences after the Great War.[11]

Both self-improvement and the consumer cultures exemplified by *Joy* were characteristic of the 1920s. If cinemagoing and celebrity are familiar historiographical tropes, there has been little work on the significance of self-improvement. New forms of popular leisure anticipate central strands in the formation of British modernities; the Standard Art Company's correspondence courses now seem anachronistic. The Pelman Institute has figured in analysis of correspondence schools, the democratization of writing, and new forms of spirituality. The most sustained discussion treats self-improvement as exemplary of practical psychology. Within the history of ideas, Steffan Blayney and Mathew Thomson use Pelmanism, which emphasizes individuals' ability to harness the potential of the self within, to trace the diffusion of modes of psychological knowledge. This approach means we have lost sight of self-improvement's affinities with wider patterns of society and culture. Isolating self-improvement as a manifestation of practical psychology makes it harder to see its contemporary resonance and its significance for our understanding of the period as historians.[12]

The divergence of these historiographies is instructive: a focus on shifting psychological ideas clashes with the burgeoning literature on mass culture, social change, and personhood. This bifurcation reflects how cultural forms were shaped by differences of class, gender, and age. Advertising and pamphlets associated self-improvement with the public realm of work and commerce and with an aspiring, earnest, status-conscious middle class. *Joy*, by contrast, exemplified commercial cultures that critics often dismissively linked to the fantasies of young working-class women. Correspondence courses were most prominent in middlebrow newspapers and periodicals rather than in story papers characteristic of mass culture. The distinctions were never absolute, but they exercised a powerful influence on how contemporaries understood different forms and how they have been interrogated by historians.

This chapter, by contrast, brings self-improvement and character competitions into the same analytic frame. These forms, I argue, pivoted around common problems of identification and addressed the fraught relationship between appearance and character through shared logics and technologies. Viewing them together suggests new ways of understanding ostensibly discrete cultural forms. More importantly, it allows us to think critically about the 1920s as a particular historical conjuncture. What did newspaper competitions and self-improvement have in common, and why did they become prominent at the same time? To address these questions, I explore what happens when we make two intellectual moves. First, we treat self-improvement as analogous to fantasies of social advancement usually associated with cinema fandom or escapist fiction. Second, we treat competitions and stunts as versions of self-improving projects usually associated with the autodidact. Approached in this way, dilemmas of identification underscore the tensions between ideas of personhood, new cultural forms, and differences of class and gender.

This means comparing courses peddled by ABC Correspondence Schools with the kind of competitions Pierres entered. Novello worked from photographs, but in a "society of strangers," judging character was both mundane and freighted, and attentiveness to the minutiae of self-presentation was embedded in everyday social and economic relations.[13] Courses such as "How to Size People Up from Their Looks" sought to profit by promising to equip citizens for modern life; competitions provided opportunities to practice those skills. In each mode, identification was notionally a systematic and sequential process, unfolding as discrete steps culminating in the moment when someone could be placed within a typology of character. The internal logics of that process might differ, but to work, it always required an end point that was fixed and immutable—a notion of character as innate and subject to appraisal.

In the 1920s, however, that notion was under sustained pressure. The experience of the Great War and the discussions of psychological breakdown that coalesced in the 1922 War Office Committee of Enquiry into Shell Shock suggested that modern war rendered ideas of character predicated on manliness and self-control untenable.[14] Psychoanalytic ideas of selfhood and Einstein's theory of relativity suggested that, as contemporary historians Robert Graves and Alan Hodge have observed, "people were no longer people but merely peripatetic points of view."[15] The guiding principles of "sizing up" were called into question. The treatment by Charming Faces of character as interior and identifiable existed alongside growing emphasis on personality as an external surface that could be refashioned. Advertising for the beauty industry traded in the possibilities of transcendent social mobility: regardless of the realities of class or wealth, self-will and judicious choice of commodities could bring success in life and love. Commercial culture reconfigured social identities around ideas of performance or style, just as correspondence courses promised the personal transformation and social advancement that would follow from learning the simple tricks necessary to secure confidence. Following work on the United States, I position self-improvement within this wider terrain. Those "fables of abundance" played out in the work of entrepreneurs such as Dale Carnegie interwove ideas of personal efficiency and the managed self into a commercialized "perfectionist project" that cut across mass and middlebrow cultures.[16]

The problem of identification thus played out on two planes. It crystallized first as careful scrutiny of photographs by intrigued competition entrants and earnest self-improvers. That scrutiny also marked a crisis, however—something we might understand as the impossibility of identification in a plausible world. Character competitions and self-improvement mobilized ideas of personhood that disrupted the connections between internal character and external appearance. How could the "truth" of character be isolated when personality was mutable, devolving into a set of surfaces that might be refashioned to get ahead? Personality was an ongoing process, encapsulated in the deliberate work of self-improvement and self-presentation. This shadowed the logics of identification but disrupted the sequence and object of that process. As the elaborate infrastructure sustaining the business and pleasure of identification expanded in the 1920s, its end point became a shibboleth. The result was troubling: locating individuals within typologies of character was a compelling challenge yet increasingly difficult. Placing something as mundane as Pierres's certificate within the conditions that made it possible allows us to understand mass society's intractable problem of identification. As something to think with, Charming Faces reveals as much as the canon of modernist literature.

Judging

Charming Faces exemplifies the stunts, giveaways, and competitions through which newspapers and magazines competed for readers and advertisers in the 1920s. Shaped by intense economic pressures, it also betrayed pervasive interest in the problem of identification. Periodicals regularly invited readers to submit photographs for analysis or to pass judgment on others. Being scrutinized by Novello was unusual; the underlying logic was ubiquitous: personality could be gauged through nuances of physiognomy, expression, and appearance. Snap judgments about strangers were the weft of everyday life. When the *Sunday News* launched a £100-prize competition for reading "Character from Faces" in August 1926, they characterized this process as "instinctive." Readers were asked to "throw your mind back to your last holiday, when you sat with your arms round your knees on the sands and drawled out comments on the people who strolled idly by. 'She's attractive—hmm—but she has a devilish temper!'" Assessing character also required diligent practice, however. Balancing entertainment and education, the paper offered holidaymakers an amusing diversion that equipped them for modern life.[17]

Newspaper competitions were not alone in addressing the problem of character. After the Great War, a growing number of schemes presented analysis of the body's external surfaces as a science to learn or skill to nurture. Isolated as an intriguing problem, the business and pleasure of identification were sustained through commercial networks that collapsed the distinctions of form and brow and the differences of class, age, and gender. *Joy*'s readers were young and female, but Charming Faces echoed self-improvement programs aimed at older, wealthier men. Entrepreneurial boosters sought to exploit economic dislocation and social uncertainty and to mobilize the anxieties of a beleaguered middle class; the correspondence courses of the Pelman Institute and ABC Correspondence Schools addressed profitable demand for instruction and advancement. Their portfolio encompassed professional skills and accomplishments such as life drawing, piano playing, and journalism. It also included ambitious programs of individual transformation such as "Mr Purinton's Personal Efficiency Course."[18] All of these projects shared the assumption that cultivating personality could bring personal and professional success. Presented as an alchemic project and advertised in publications such as the *Daily Mail* and *Country Life*, self-improvement sold the demotic possibility of social mobility achievable by developing untapped "potential" within.[19]

Exemplified by Dr. Katherine Blackford's National Business and Personal Efficiency Programme, this version of self-improvement focused on three entwined strands: memory, self-presentation, and character appraisal. In seeking to give students the skills necessary for professional and commercial success, Blackford's pamphlets were the leading edge of the new science of human relations in the United States and reflected the Standard Art Company's extensive market in transatlantic reprints. This was a commercialized traffic in ideas through which courses were repackaged for British readers: those new modes of personhood associated with self-improvement were part of a wider process of Americanization.

Like other courses, Blackford's *Judging Character at Sight* (1919) collapsed social advancement into the ability to "size people up from their looks." Seven lessons taught the "little signs that read character at a glance." Hidden from casual observers, these were as "plain as print" for the initiate. Overblown testimonials explained the payoff:

> I can now tell almost the minute I lay eyes on people how to make them my friends, in either a business or social way—how to talk to them; how to influence them to the best advantage. Also I can tell at a glance whom I can trust and whom I can't.[20]

There was a magical element to self-improvement—a kind of daydreaming, tempered by the reason and rigor necessary to get ahead. It was a fantasy, of course, but the idea that a searching gaze might be one of modern life's "most valuable assets" was a powerful selling point.[21]

Judging Character was underpinned by three principles. First, correspondence courses had transcendental potential. Mobilizing aspiration and anxiety, Blackford described how many were "not only unsuccessful in business or profession, but unhappy in family life, and starved for companionship and friendship."[22] Rather than a structural critique of inequality, this was a reason to invest in individual self-improvement. Diligent study would ensure "failures become successful—the unpopular becomes popular." More than this—judging character would bring the "ability to make people like you, to make them believe you, and to get them to do as you wish."[23] Second, physiognomy and character aligned. Personal attributes could be mapped onto specific facial features. Students began by scrutinizing their profile with the aid of a mirror, camera, and annotated diagrams and photographs. Having situated themselves within a typology of "convex," "concave," and "plane," they could then use a table to read across from classification to character.[24] Methodical and deliberate, this process underscored the third principle: here was a "simple *scientific* formula for achievement."[25] Starting with self-analysis meant that "you will know, when you come to the description of your profile and the traits which go with it, that you are reading something scientific and authoritative."²⁶ Blackford grounded her typologies in a survey of eighteen thousand individuals. In this guise, identification was a rigorous, rational "system" rooted in "years of painstaking analysis." In a claim reminiscent of contemporary criminology, Blackford and others asserted themselves to be privileged guides to modern life.²⁷

Organizations such as the Pelman Institute focused on the public world of work, addressing the man of business and evoking an austere masculine personhood rooted in determined self-will. While these schemes were ostensibly distanced from mass cultural forms such as *Joy*, the resonance between self-improvement and competition stunt went beyond dilemmas of identification. The *Sunday News* promised to entertain readers; Blackford noted how "tackl[ing] the baffling problems of personality with scientific honesty and human sympathy" would bring "new interest and pleasure in life." Advancement went alongside emotional "satisfaction" and leisure.²⁸ Blackford showed the apparatus of camera and mirror needed for self-analysis; the magazine *Popular* introduced their "Character from Photograph" competition with a sketch of a woman contemplating her reflection in a mirror. As she brushed her hair, her intent gaze embodied *Judging Character*'s invitation to self-scrutiny.²⁹

The *Sunday News* competition demonstrated this movement of ideas. Ventriloquizing Blackford's advertising, the judge exhorted readers to "search faces and succeed!" Learning to read the "ordnance map" of feature and expression would unlock the "mysteries" of character. This searching gaze was essential for the colonial administrator or businessman but also had far-reaching utility. Reading character could be a necessary emotional defense: countless "unhappy marriages" might be averted if a charming stranger was given away by their close-set eyes, which "betray[ed] the vile temper he sought to camouflage by a particularly attractive smile." When colonial subjects, prospective employees, and suitors could all be acting, such skills afforded a hardy self-reliance, penetrating the "plausible effusiveness" that made life a confidence game.³⁰

That the artifice of a promotional stunt could provide training for modern life underscored the reach of the business and pleasure of identification. The "fortune in face reading" denoted *both* the "valuable prize [on offer] to the skillful face reader" and how such skills could bring success in "love and business." Here was Blackford's transcendental promise—the science of human relations, played out through a weekly competition.³¹ Character analyses in *Picturegoer* invoked the "principles of physiognomy" that defined the "relationship between feature and character."³² In *Popular*, Stylo offered methodical analysis of the meaning of facial features. The absence of published reader photographs (despite repeated requests) suggests the stunt was unsuccessful. Isolating the eyes or chin for scrutiny still provided building blocks for identification.³³

The rhythm of such journalistic features also echoed correspondence courses. *Judging Character* was a series of didactic lessons and practical exercises. Reminiscent of newspaper competitions, photographs tested students' grasp of physiognomical theories: "Write your reading of the following characters in the space indicated."³⁴ The *Sunday News* posed similar challenges. Presented with a cropped studio portrait—carefully posed, features often highlighted in ink—readers were asked: "Imagine you had just met the owner, what would your unspoken verdict be?" A table listed twelve personality attributes: Was a smart woman "self-willed" or "sanguine"?³⁵ Was the mustachioed man "ambitious" or "abstemious"?³⁶ The reader who ticked the right boxes could win a prize.³⁷ Problems of identification could be addressed for entertainment, education, and profit; a hands-on process cultivated the "face-reading habit."³⁸

The parallels went further. Pelman students received individual feedback; *Sunday News* readers read expert commentary on competition results, which "reveal[ed] the secrets of facial lines and . . . assist[ed] you in your reading of the picture we publish[ed] this week." Detailed, often caustic, they demonstrated "face-reading" in action. One photograph posed a dilemma: how could you tell its subject was suave and avaricious? The columnist explained:

> His utter want of reliability . . . is betrayed by the big space from the top of his forehead to the bridge of his nose, while his suavity . . . is shown by the shape of his heavy-lidded eyes, whose forced expression of frankness . . . conceals an overwhelming avariciousness.³⁹

This was someone to beware, not believe. Rooted in the serials used to hook readers, the repetitive competition form also manifested self-improvement's pervasive influence.

When the *Sunday News* reflected on the man with heavy-lidded eyes, they noted that the accomplished "face diviner would realise that the face in the picture is not a normal or healthy minded one." Such confidence belied the uncertainties of judging character. Blackford and others gestured toward the research informing their systems, and competitions deployed the scientific language of physiognomy. There was little sense of the systemic knowledge behind their assessments, however.[40] Stylo traded in normative assertions, not general principles. The claim that "insignificant noses denote insignificant people" did not define significance, let alone explain the meaning of a wider or narrower nose.[41]

The possibility of identification was thus unsettled: how could anyone be sure the *Sunday News* was right? Mug shots, fingerprints, and biographies compiled by the Criminal Records Office secured authority through a circular process: successful identification of an individual verified the integrity of the system. Newspaper competitions, by contrast, compromised their claimed authority. The correlation between features and character was imprecise; appearance was elusive. A woman's face showed "divers characteristics which ... contradict each other in a bewildering way." In the absence of agreed-on principles or method, identification depended on instinct or prejudice. The "shape of her eyelids" showed this woman was "loveable," but the overall conclusion was damning: she was impressionable ("low forehead") and morbid ("size her of eyeballs"). This analysis did not explain how to weigh eyelid shape against eyeball size.[42] Such challenges were compounded by those trying to evade scrutiny. Another face—"changeable" and "deceptive"—was "remarkable for its misguiding mask of beauty." Identification could never be more than provisional, and Britons had to beware the masks behind which strangers might hide.[43]

These problems were compounded by the dubious authority of those passing judgment. "Practice," *Joy* suggested, "made [Novello] the quickest and surest reckoner-up of modern girls in England."[44] His qualifications were fame, not formal training; celebrity, not careful study; and a roving eye, not reason. Stylo was anonymous, but the nom de plume suggested a seaside magician rather than an expert in human relations. The *Sunday News* competition was introduced not by a criminologist but by "Miss Nell St John Montague, the famous Society clairvoyant." Montague never explicitly invoked the psychic as a source of authority. Despite the jargon of scientific analysis, the foundations of her knowledge were uncertain. Identification thus depended on the clairvoyant's supernatural capacities and aristocrat's privilege rather than on specialist knowledge.[45]

In 1924 "John Blunt" observed how "everybody's opinion of everybody else is . . . largely formed from a subconscious impression created by the face." If the connection between physiognomy and personality was commonplace, being quoted in an article on cosmetics in *Picturegoer* was paradoxical.[46] Like other cinema magazines, *Picturegoer* regularly published character analyses. A series by American journalist Vincent de Sola profiled stars such as Anita Stewart from photographs. Yet de Sola's byline as a "character expert" belied how his column exposed the contradictions of judging character.[47] He never acknowledged the difficulties of gauging personality from production stills and publicity images defined by artifice and stylized glamour. The headshots illustrating columns such as de Sola's and *Picture Show*'s "Character as Told by the Face" were, moreover, identical to another established feature—"The Expressions of . . ."[48] Captioned photographs, cropped to magnify details of feature and expression, showed Priscilla Deane and others in different moods: "Charming and Coy" or "So Sad." Despite a shared visual economy, the logic of these columns diverged. Rather than providing a basis for isolating the "truth" of character, photographs celebrated the ability to create illusions of emotion through facial dexterity and technique.[49]

"Expressions" reflected a particular postwar moment—the growth of cinemagoing and celebrity culture and of interest in the relationship between appearance and character. Framed as a guide to the method of acting, it exemplified a demotic aspirational culture and the proliferating institutions promising to turn ordinary Britons into stars. As Chris O'Rourke shows, this didactic apparatus encompassed cinema schools, "star search" competitions, correspondence courses such as *Cinema Acting as a Profession* (1919), and insider guides such as Fred Dangerfield and Norman Howard's *How to Become a Film Artiste* (1921). Actors took readers behind the scenes of screen performances; journalists shared tales of the typist turned starlet.[50]

"Expressions" also embodied a specific version of acting as a craft. The demands of silent cinema technologies and contemporary notions of emotional veracity placed a premium on facial expressions. In this guise, acting was understood as a process of "registering" emotions on screen. "Try it yourself in front of a looking-glass," challenged *Picture Show*. "It is so difficult not to grimace or exaggerate."[51] Done properly, mobilizing movements of mouth or eyes conveyed mood and secured audience trust. *How to Become a Film*

Artiste stressed the "importance of portraying character and emotion primarily through gesture and facial expression" and provided exercises to develop those skills.⁵² The "story of the film," it counseled, should be "told by natural movements of the eyes, hands, lips . . . in accord with the emotions to be portrayed." It was difficult, though, to reconcile the injunction that movements should be "natural" with the idea that acting was a craft to hone.⁵³

The "accord" of expression and emotion was a challenging ideal. *Picture Show* praised Pauline Frederick because "her face mirrors her thoughts and emotions in a most expressive way." It was Frederick's "faculty for showing what she feels" that made her "one of the greatest stars today." Such comments secured the connection between physiognomy and personality: if bodily surfaces were a "mirror" for internal moods, Frederick was feeling, not performing, on-screen.⁵⁴ At the same time, criticism evinced competing visions of acting. Irene Rich had "one of the most valuable assets a screen player can possess—the ability to portray emotion though facial expression." Emotion and expression diverged: if acting was the ability to create an ersatz facsimile, gauging character became even harder.⁵⁵

Such journalistic features emblematized wider interest in the masquerades shaping social relations and ideas of personhood. "Popular interest in film acting," notes O'Rourke, "coincided with the notion . . . that individuals needed to cultivate the skills of the actor in everyday life in order to be successful."⁵⁶ *Joy* described being typified as revealing "the part Ivor Novello would allot you in the drama of life."⁵⁷ Advertising traded in ideas of personality as self-presentation—a mutable exterior that could be refashioned rather than a function of interior character. Mass culture again collapsed into self-improvement. Correspondence courses taught Britons how to act and trained them, through courses such as Modern Salesmanship, in the performances necessary for social advancement.⁵⁸ ABC Correspondence Schools played on anxieties to sell courses in public speaking, warning that "KEEN COMPETITION is at hand. How do you propose to insure yourself against a falling off in your income? How do you think you are going to *increase* your income?"⁵⁹ Testimonials endorsed the magical effects: "Correct speech enhances one's social attractions—people are ever ready to listen to an interesting and engaging talker."⁶⁰

What the Pelman Institute termed the "science of Self Realisation" was thus intertwined with and reminiscent of dreamlike fantasies of social mobility usually associated with cinema fandom; a hard-headed fixation on career and financial gain became a more amorphous pursuit of happiness or fulfillment. Consider the institute's promises: "Rise to a higher position . . . , [and] increase your income-earning power."[61] Building self-confidence was integral to Pelmanism. If this could be achieved by addressing "defects and weaknesses" and cultivating "positive, vital qualities," success was measurable only through responses. Cultivating self-belief turned into teaching individuals how to mobilize the confidence of others—to "become a clever salesman . . . , acquire a strong personality . . . , [and] win the confidence of others."[62] Such qualities reduced social relations to what Graves and Hodge called a game of "poker-play technique and exercise of personality."[63] Self-reflection, Blackford counseled, allowed individuals to identify "points which make your personality attractive" and those "which detract from the power of your personality." Self-knowledge enabled experts to "develop and make the most of your strong points" and at the same time "keep your weak points in the background."[64] Self-realization transformed into the quotidian performances through which others could be manipulated. Advertising and advice literature made petty deceptions and bold claims integral to everyday life.

The language of self-realization moved from and between correspondence courses and competitions, film magazines and fan letters, and commercial advertising. Consider Amami shampoo's extensive publicity: the famous slogan "Friday night is Amami night" exemplified the promotion of beauty products as affording new possibilities for self-fashioning. Endorsed by fashionable icons, such products acquired alchemic properties that would give users the chance to realize frustrated potential and become "as glorious as Theda Bara."[65] Amami supported Charming Faces, offering prizes to winners who proved they were aficionados by submitting an empty sachet.[66] Later they celebrated "another Amami success" with a full-page portrait of *Joy*'s winner Norah Horan, who attributed her hair's "perfect condition" to "regularly shampooing with Amami."[67] This was the context for the "Amami Film Star Quest." Cosponsored by British International Pictures, the search for a "Great British Film Star" sought to reinvigorate national cultural production. Newspaper stunts, celebrity endorsements, and branding intersected around the possibilities of personal transformation. The winner, Eugenie Prescott, was symbolically renamed.[68] Exploiting her newfound fame, further campaigns presented readers with a series of photographs and a question: "What emotions are being expressed by Eugenie Amami?" Those who could distinguish "disbelief" from "teasing" stood to win a pair of stockings.[69] The competition suggested identity was self-evident yet sat alongside adverts that cultivated aspirational fantasies of self-improvement.

Magnetism

In identifying Stella Pierres as the "magnetic type," Ivor Novello evoked both an everyday correlation between physiognomy and personality and the interlocking practices of identification that underpinned self-improvement and popular print culture. Here, "magnetism" was a recurring motif. Vincent de Sola noted the "striking sex-magnetism" in Gloria Swanson's portrait, isolating the "lucent eyes, set with a slightly oblique lift" as "indications of one who will always attract love." Novello and de Sola might have disagreed over the signs betraying character; de Sola acknowledged a "highly mutable" expression that meant Swanson's "face is by no means easy to read." Magnetism was still treated as an attribute that trained observers could locate in features or demeanor.[70]

Magnetism and its corollaries "charm" and "appeal" focused the problem of identification. It was never clear if these were qualities to possess, tactics to deploy, attributes inhering in an individual, or effects materializing in relationships with others. De Sola saw magnetism in Swanson's eyes yet noted the "practicality" that "enabled her to employ her magnetic appeal ... profitably."[71] That magnetism might be deployed reflected the promise of self-improvement. If personality could be developed, ordinary citizens could gain the "magnetic appeal" necessary to influence others. A publicity sketch for Blackford imagined the possibilities: after dinner, perhaps, a fashionable man held court, his dazzling personality manifested in poise and posture, with three glamorous women captivated by his charm. Materialized in the scene's composition, magnetism was an irresistible force drawing in those around the table. Such powers, which were essential to social and sexual success, were accessible to all. Blackford promised to share the "secrets of being able to hold people's interest ... and become a popular idol wherever you go."[72]

That appeal could be acquired belied both the notion of magnetism as a "type" and the courses offering to teach character judgment. Setting out the skills necessary to succeed in public life, Frederick Houk Law's *Mastery of Speech* identified the "psychological foundations of speech—the art of making personal appeal to your hearers."[73] Law's version of "appeal" was a lesson in the everyday practices of magnetism. "Whether with one person or a thousand," he counseled, "look into the eyes of your hearers. Let your personality meet theirs." Eye contact, performative listening, compliments, and conversational common ground—these were tricks to cultivate confidence and evade scrutiny.[74] Exercises provided a blueprint that readers could use to "establish a magnetic bond that will not easily be broken." Law sometimes suggested that "personality" was a bounded entity, but more often, magnetism was a game of manipulating others.[75]

Like the literature of human relations, *Mastery of Speech* focused on the masculine realm of business. The separation of commerce and mass culture was a self-serving fiction, however, and Law's strictures were echoed in cinema and fashion magazines. In *Picturegoer*, June weighed the merits of "charm" and "beauty." Social and sexual success, she argued, depended on appeal, not appearance. This made "magnetism and personality . . . things to cultivate" to get ahead. Like Law's students, readers received a how-to on becoming magnetic:

> Cultivate a pretty manner, a low pitched voice, and you will find people listen to you with interest. . . .
> Speak slowly. . . . There is such a thing as a well poised voice, and poise . . . is a most important thing. Cultivate an air of assurance whether you feel sure of yourself or whether you do not.

Scientific self-improvement was reworked as advice for young women. Rather than a type, magnetism was a demanding process of self-scrutiny and management. "Watch for and weed out any unpleasing mannerism," June advised. "Even our emotions are sometimes pitched a tone or so too high," so "we are apt to get hysterical and over-excited and then goodbye charm." Social success was a demotic goal but required effortful work. June concluded, "Charm, that subtly magnetic thing that attracts any and everybody to you whether they will or no, is worth all the beauty secrets ever invented."[76]

"Whether they will or no"—here self-improvement has moved from cultivating personality to securing compliance in others. Courses sharing the secrets of judging character also presented getting ahead as a confidence game. Law and June taught everyday arts of masking that precluded identification, or at least disrupted its logics. Contemporaries acknowledged these contradictions. Entrepreneurs traded in personal transformation but insisted this meant realizing the self within rather than creating a front to hide behind. *Judging Character* clumsily addressed the problems caused by "egoists and bluffers," saying that while the "clever actor... may fool even the shrewdest for a time," most of those who "try to assume what they do not really feel, overdo it." Here was reassurance: "Watch them closely, and their real class will show in little ways they do not suspect."[77] This defensiveness suggests how educationalists and journalists were preoccupied with the difficulty of distinguishing "bluffers" from the "real thing" and stabilizing the relationship between personhood and its external expression. Yet "magnetic appeal" or a "charming face" denoted irreconcilable understandings of the relationship between appearance and character, individual and society, self and other. How could it be possible to ascertain the "truth" of identity when courses taught "the secret of being a convincing talker"?[78]

Conclusion

Like Ivor Novello, Virginia Woolf struggled with problems of identification. Presented with hundreds of photographs, Novello had to assign each "charming face" to a character type. Similarly, in contemplating the "myriad of impressions" received by the "ordinary mind," the novelist had to give form to an "incessant shower of innumerable atoms."[79] "An Unwritten Novel" (1920) explored how storytelling might be a process of judging from such impressions. Rather than concentrating on a cropped photograph, Woolf focused on the face framed by her newspaper and press of passengers on a busy train. There was "knowledge in each face," Woolf noted, though most tried to "hide or stultify" it behind a book or cigarette. Fixating on an "unfortunate woman" who could not "play the game," Woolf's intrusive gaze translated the minutiae of expression into sadness, perceived tragedy into imagined familial melodrama. "Leaning back in my corner, shielding my eyes from her eyes..., I read her message, deciphered her secret." The unfolding story was punctured by moments of its making, of "peeps" across the carriage, as though seeking inspiration. Despite doubts—"Have I read you right?... The human face... holds more, withholds more."—by the time they alighted, Woolf was confident: "Though we keep up pretences, I've read you right—I'm with you now."[80]

We are used to thinking about the problem of identification as a challenge for state bureaucracies and national security. That is right, of course—fraught debates around citizenship and crime punctuated the postwar decade. Yet Woolf's and Novello's dilemmas as well as the competitions and courses explored here suggest that the problem also animated networks of business and pleasure that collapsed boundaries between state, society, and culture and moved across putative distinctions between cultural forms and expressions. Bringing self-improvement, journalistic stunts, and public relations into the same frame suggests how the infrastructure of commercial mass culture carried shifting ideas of personhood. The proliferation of the tight-framed portrait photograph in passports, police records, character competitions, and correspondence courses was no coincidence. There is a simple story here about how criminological or anthropological modes of identification could bleed into high, middlebrow, and popular culture. In simply noting this movement of ideas, however, we reduce cultural history to description and evade the more important task of understanding *why* that traffic occurred and *what* it can tell us about 1920s Britain. Here is a more challenging story of how Charming Faces marked a crisis of identification.

Thinking historically about self-improvement and character competitions means understanding their contemporaneous prominence and formative conditions. These tightly imbricated forms reflected a moment—roughly between the end of the Great War and the late 1920s—in which dilemmas of identification were distinctive and particularly intense. Both forms were underpinned by a flourishing popular print culture in that brief period before competition from newsreels, cinema, wireless, and television challenged its centrality in commercial culture. The growth of cinemagoing and of feature films and celebrity cultures that elicited intensified emotional investments from audiences joined with technologies for making and exhibiting silent films to shape the exaggerated performance styles associated with "registering" and secured the connections between expression and character. Animated by characteristic networks of commercial culture, self-improvement and character competitions traded in personal transformation, holding out the prospect of social mobility unconstrained by class or wealth. In an uncertain postwar world, they reflected an obsession with the problematic relationship between inside and outside and the difficulties of interpersonal relations between strangers.

In one sense, these modern forms of self-improvement echoed Victorian ideals associated with the work of Samuel Smiles. *Self Help* (1859) and *Character* (1871) emphasized how diligence, thrift, and hard work could bring advancement; Smiles stressed the virtues necessary for material success. Later schemes increasingly presented self-realization as the cultivation of hidden psychological potential and of the capacities of the self within. The distinction was never clear-cut but crystallized as a slippage from character to personality. Drawing on the language of popular psychology, postwar self-improvement commodified Victorian traditions of self-help, exploiting the technologies provided by the modern mass newspaper and postal network. That it still paid lip service to the difficult work of reason and reflection was belied by ads presenting self-improvement as a commodity to buy or as the magical key to social success. Above all, perhaps, the distinctiveness of the 1920s was encapsulated in the concurrence of incompatible ideas that character could be read from appearance, while personality could be refashioned.

This moment was fleeting. While commercial self-improvement continued, it lost impetus starting in the mid-1920s. A marked decline in correspondence school advertising reflected particular financial difficulties—for example, the Pelman Institute went bankrupt in 1921. It also reflected the stabilization of middle-class incomes. After a short postwar crisis, economic recovery and social stability attenuated the insecurities, which entrepreneurs had exploited. Character competitions also became less prominent in the late 1920s. Finally, the introduction of sound into cinema production and exhibition changed the nature of acting. The obsession with facial expressions associated with registering softened, just as formalized routes into the profession reduced opportunities for aspiring actors and made dramatic social mobility less imaginable.

Postwar self-improvement anticipated the spectacular development of our contemporary happiness industry. Like that industry, self-improvement traded in the demotic illusion that material realities could be transcended through strength of mind alone, even as they remained individualistic in orientation and underpinned by deep-rooted privileges of class and wealth. In this fictive mode, social mobility and well-being were deemed entirely functions of personality.[81] Self-improvement's utopian tendencies nonetheless marked a pressing problem of identification. This was where "An Unwritten Novel" ended. Woolf's sense of closeness to her subject dissolved when she witnessed the greeting before her. Uncertainty—"there's something queer in her cloak"—became the abrupt realization of the limits of knowledge. What followed was an unsettling crisis. The story's premise collapsed as Woolf confronted the impossibility of judging character: "Well, my world's done for! What do I stand on? What do I know?" No longer contained in a railway carriage or photograph, melting into the everyday whirl, personality became mobile, mutable, and "mysterious"—cloaked.[82] The dilemmas of identification explored in this chapter were analogous to the preoccupations of modernist fiction. Yet they were also more pervasive—and, as such, more revealing of the fault lines of 1920s culture. Whether the benefits of "magnetic appeal" or a "charming face" were secured through diligent study, careful self-scrutiny, or a particular cosmetics brand, the connection between outside and inside—personality and physiognomy—had become untethered.

Notes

1. Stella Pierres, scrapbook, 1924–25, LCM/326, Bishopsgate Institute, London.
2. "Charming Faces Competition," *Joy*, 14 February 1925, 35. See also "£150 and 300 Sets of Lingerie Must Be Won," *Joy*, 14 February 1925, 36; and "Charming Faces Competition," *Joy*, 21 February 1925, 2, 35.
3. "Ivor Novello's Great Offer to British Girls," *Lancashire Daily Post*, 10 February 1925, 3; "Ivor Novello's Great Offer to British Girls," *Daily Mail*, 11 February 1925, 5.
4. Ibid.
5. "Charming Faces Competition," *Joy*, 14 February 1925, 35.
6. See, for example, "A Few Photographs Chosen at Random from the Charming Faces Competition," *Joy*, 25 April 1925, 18; "First Prize Winners in the Charm Contest," *Joy*, 23 May 1925, 2; and "Three Hundred Prize Winners," *Joy*, 30 May 1925, 2, 34–35.
7. Peter Cryle and Elizabeth Stephens, *Normality: A Critical Genealogy* (Chicago: University of Chicago Press, 2017), chap. 8.
8. Michael Saler, "'Clap If You Believe in Sherlock Holmes': Mass Culture and the Re-enchantment of Modernity, c. 1890–1940," *Historical Journal* 46, no. 3 (2003): 620.
9. Ruth Livesey, "Reading for Character: Women Social Reformers and Narratives of the Urban Poor in Late Victorian and Edwardian London," *Journal of Victorian Culture* 9, no. 1 (2004); Stefan Collini, "The Idea of Character: Private Habits and Public Virtues," in *Public Moralists: Political Thought and Intellectual Life in Britain, 1850–1930* (Oxford, UK: Clarendon, 1991), 91–118.
10. Suzanne Bailey, "Francis Galton's Face Project: Morphing the Victorian Human," *Photography and Culture* 5, no. 2 (2012): 189–214; Jennifer Green-Lewis, *Framing the Victorians: Photography and the Culture of Realism* (Ithaca: Cornell University Press, 1996); Alejandra Bronfman, "The Allure of Technology: Photographs, Statistics and the Elusive Female Criminal in 1930s Cuba," *Gender and History* 19, no. 1 (2007): 68; Julia Laite, *Common Prostitutes and Ordinary Citizens: Commercial Sex in London, 1885–1960* (Basingstoke, UK: Palgrave Macmillan, 2012), chap. 6; Charles Bell, *Essays on the Anatomy and Philosophy of Expression* (London: John Murray, 1824); Charles Darwin, *The Expression of Emotion in Man and Animals* (London: John Murray, 1872). For recent interdisciplinary overviews of these debates, see, for example, Paul Deslandes, *The Culture of Male Beauty in Britain: From the First Photographs to David Beckham* (Chicago: University of Chicago Press, forthcoming 2021); and Heather Widdows, *Perfect Me: Beauty as an Ethical Ideal* (Princeton: Princeton University Press, 2018).
11. Shani D'Cruze, "'Dad's Back': Mapping Masculinities, Moralities and the Law in the Novels of Margery Allingham,"

Cultural and Social History 1, no. 3 (2004); Victoria Stewart, *Crime Writing in Interwar Britain: Fact and Fiction in the Golden Age* (Cambridge: Cambridge University Press, 2017).

12. Mathew Thomson, *Psychological Subjects: Identity, Culture, and Health in Twentieth Century Britain* (Oxford: Oxford University Press, 2006); Steffan Blayney, "Health and Efficiency: Fatigue, the Science of Work, and the Working Body, c. 1870–1939" (PhD diss., University of London, 2018). See also Christopher Hilliard, *To Exercise Our Talents: The Democratization of Writing in Britain* (Cambridge, MA: Harvard University Press, 2006), 104; and Callum Brown, *Religion and Society in Twentieth Century Britain* (London: Routledge, 2006), 102–3, 199. On earlier traditions, see Malcolm Chase, "'An Overpowering Itch for Writing': R.K. Philp, John Denman, and the Culture of Self Improvement," *English Historical Review* 133, no. 463 (2018): 351–82.

13. James Vernon, *Distant Strangers: How Britain Became Modern* (Berkeley: University of California Press, 2014); Peter Bailey, "White Collars, Gray Lives? The Lower Middle Class Revisited," *Journal of British Studies* 38, no. 3 (1999): 289.

14. Ted Bogacz, "War Neurosis and Cultural Change in England, 1914–22: The Work of the War Office Committee of Enquiry into 'Shell-Shock,'" *Journal of Contemporary History* 24, no. 2 (1989): 227–56; Michael Roper, "Between Manliness and Masculinity: The 'War Generation' and the Psychology of Fear in Britain, 1914–1950," *Journal of British Studies* 44, no. 2 (2005): 343–62.

15. Robert Graves and Alan Hodge, *The Long Week-End: A Social History of Great Britain* (London: Reader's Union, 1931), 95; Katy Price, *Loving Faster than Light: Romance and Readers in Einstein's Universe* (Chicago: University of Chicago Press, 2012).

16. Jackson Lears, *Fables of Abundance: A Cultural History of Advertising in America* (New York: Basic Books, 1994); Kathy Peiss, *Hope in a Jar: The Making of America's Beauty Culture* (Philadelphia: University of Pennsylvania Press, 2011); Edwin Battistella, *Do You Make These Mistakes in English? The Story of Sherwin Cody's Famous Language School* (Oxford: Oxford University Press, 2009); Steven Watts, *Self-Help Messiah: Dale Carnegie and Success in Modern America* (New York: Other Press, 2013); Trysh Travis, *The Language of the Heart: A Cultural History of the Recovery Movement from Alcoholics Anonymous to Oprah Winfrey* (Chapel Hill: University of North Carolina Press, 2013).

17. "£100 Hidden in Lines on Man's Face," *Sunday News*, 22 August 1926, 4.

18. "How I Improved My Memory in One Evening!," *Sovereign*, November 1919, 81; "How to Master the Piano in Three Months," *Sovereign*, January 1921, ii; "The Secret of Being a Convincing Talker," *John Bull*, 21 June 1919, 15.

19. Contemporary discussions can be found in Graves and Hodge, *Long Week-End*, 188–89; and John Collier and Ian Lang, *Just the Other Day: An Informal History of Great Britain Since the Great War* (London: Hamish Hamilton, 1930), 29.

20. "How to Size People Up from Their Looks," *John Bull*, 31 May 1919, 15.

21. "Character vs. Credit," *Sovereign*, December 1919, xvi.

22. Katherine Blackford, *Judging Character at Sight* (London: Standard Art Book, 1919), 3.

23. Ibid., 4.

24. Ibid., 1.

25. Ibid., 1, 5 (emphasis mine).

26. Ibid., 7.

27. "Character vs. Credit," *Sovereign*, December 1919, xvi.

28. Blackford, *Judging Character*, 52.

29. "Character from Photograph," *Popular*, 7 March 1925, 136.

30. "There's Fortune in Faces for You!," *Sunday News*, 8 August 1926, 3.

31. Ibid.

32. "Lois Laughs at Men," *Picturegoer*, April 1925, 23.

33. "Your Character from Photograph," *Popular*, 14 February 1925, 43; "Character in

Faces III.—Eyes," *Popular*, 28 February 1925, 120; "What the Mouth and Lips Denote," *Popular*, 28 March 1925, 264; "Chins and Cheeks," *Popular*, 4 April 1925, 304.

34. Blackford, *Judging Character*, 38.

35. "There's £100 for You in This Page," *Sunday News*, 15 August 1926, 3.

36. "£100 a Week for a Face," *Sunday News*, 5 September 1926, 4.

37. "There's £100 for You in This Page," 3.

38. "Another £100 for You to Win," *Sunday News*, 19 September 1926, 7.

39. "Woman's Face Worth £100 to You," *Sunday News*, 29 August 1926, 4.

40. Ibid.

41. "Your Character from Photograph II.—Noses," *Popular*, 21 February 1925, 87.

42. "£100 a Week for a Face."

43. "Another £100 for You to Win," 7. See also "Have You Got That £100 Look?," *Sunday News*, 26 September 1926, 7.

44. "Charming Faces Competition," *Joy*, 14 February 1925, 35.

45. "There's Fortune in Faces for You!," 3. Montague—a pseudonym of Eleanor Lucie-Smith—was also a novelist and actor. See *The Glorious Adventure* (1922); Nell St. John Montague, *Revelations of a Society Clairvoyante* (London: Thornton Butterworth, 1926); and *Notable Personalities* (London: Whitehall, 1927).

46. "Beauty and Destiny," *Picturegoer*, April 1924, 52.

47. De Sola was a journalist specializing in popular psychology. See *A Thousand Dreams and Their Meanings* (New York: Mayflower, 1924). The series included "The Flapper's Favourite," *Picturegoer*, March 1925, 34; "Anita Stewart Is Fastidious," *Picturegoer*, June 1924, 30; and "A Woman's Woman, by de Sola," *Picturegoer*, August 1924.

48. "Character as Told by the Face," *Picture Show*, 1 May 1920, 20; "Character as Told by the Face," *Picture Show*, 22 May 1920, 21.

49. "The Expressions of Priscilla Deane," *Picture Show*, 1 May 1920, 9. See also "The Expressions of Monroe Salisbury," *Picture Show*, 22 May 1920, 7; and "The Expressions of Joyce Dearsley," *Picture Show*, 24 July 1920, 9.

50. Chris O'Rourke, *Acting for the Silent Screen: Film Actors and Aspiration Between the Wars* (London: I. B. Tauris, 2017). See also Lisa Stead *Off to the Pictures: Cinema-Going, Women's Writing, and Movie Culture in Interwar Britain* (Edinburgh: Edinburgh University Press, 2016); *Cinema Acting as a Profession* (London: Standard Art Book, 1919); and "Learn to Act for the Cinema in Your Own Home," *John Bull*, 25 October 1919, 15.

51. "The Expressions of Irene Rich," *Picture Show*, 24 May 1920, 7.

52. O'Rourke, *Acting for the Silent Screen*, 50.

53. Ibid., 57.

54. "The Expressions of Pauline Frederick," *Picture Show*, 2 October 1920, 7.

55. "Expressions of Irene Rich," 7.

56. O'Rourke, *Acting for the Silent Screen*, 10.

57. "Charming Faces Competition," *Joy*, 14 February 1925, 35.

58. "Be a Super-Salesman," *Sovereign*, October 1921, ix.

59. "15 Minutes a Day Will Make You a Brilliant TALKER," *Sovereign*, March 1921, v. See also "Your Speech Determines Your Money Value," *Sovereign*, July 1921; "Are You a Blonde?," *Sovereign*, January 1925, ii; and "Develop Your Personality," *Sovereign*, March 1925, vii.

60. "Are You a Convincing Talker?," *Sovereign*, December 1919, xi.

61. "Your Unsuspected Self," *John Bull*, 20 September 1919, 9.

62. "Sir John Foster Fraser's Appeal," *Weekly Dispatch*, 29 January 1928, 9.

63. Graves and Hodge, *Long Week-End*, 222.

64. Blackford, *Judging Character*, 15.

65. Mary Bertenshaw, *Sunrise to Sunset: An Autobiography* (Manchester, UK: Printwise 1991), 110.

66. "An Extra Cash Prize," *Joy*, 21 March 1925, 29.

67. "Another Amami Success," *Joy*, 30 May 1925, 17.

68. O'Rourke, *Acting for the Silent Screen*, 120–24.
69. "Amami Girl!," *John Bull*, 9 February 1929, 18.
70. "A Magnetic Movie Star," *Pictures and Picturegoer*, March 1924, 30.
71. Ibid.
72. "Develop Your Personality!," *Pictures and Picturegoer*, August 1925, 65; "Develop Your Personality," *Sovereign*, November 1925, v; "The Secret of Making People Like You," *Sovereign*, November 1919, 90.
73. Frederick Houk Law, *Mastery of Speech: A Course in Eight Parts on General Speech, Business Talking, and Public Speaking* (London: Standard Art Book, 1920).
74. Ibid., 51.
75. Ibid., 9.
76. "Charm," *Pictures and Picturegoer*, December 1924, 80.
77. Blackford, *Judging Character*, 76; "Seek Your Self!," *John Bull*, 14 February 1925, 9.
78. "The Secret of Being a Convincing Talker," *Sovereign*, June 1925, x.
79. Virginia Woolf, "Modern Fiction," in *Monday or Tuesday* (London: Hogarth Press, 1921).
80. Woolf, "An Unwritten Novel," in *Monday or Tuesday* (London: Hogarth Press, 1921).
81. William Davies, *The Happiness Industry: How the Government and Big Business Sold Us Well-Being* (London: Verso, 2015).
82. Woolf, "An Unwritten Novel."

References

Anon. "Amami Girl!" *John Bull*, 9 February 1929, 18.
———. "Anita Stewart Is Fastidious." *Picturegoer*, June 1924, 30.
———. "Another Amami Success." *Joy*, 30 May 1925, 17.
———. "Another £100 for You to Win." *Sunday News*, 19 September 1926, 7.
———. "Are You a Blonde?" *Sovereign*, January 1925, ii.
———. "Are You a Convincing Talker?" *Sovereign*, December 1919, xi.
———. "Be a Super-Salesman." *Sovereign*, October 1921, ix.
———. "Beauty and Destiny." *Picturegoer*, April 1924, 52.
———. "Character as Told by the Face." *Picture Show*, 1 May 1920, 20.
———. "Character as Told by the Face." *Picture Show*, 22 May 1920, 21.
———. "Character from Photograph." *Popular*, 7 March 1925, 136.
———. "Character in Faces III.—Eyes." *Popular*, 28 February 1925, 120.
———. "Character vs. Credit." *Sovereign*, December 1919, xvi.
———. "Charm." *Pictures and Picturegoer*, December 1924, 80.
———. "Charming Faces Competition." *Joy*, 14 February 1925, 35.
———. "Charming Faces Competition." *Joy*, 21 February 1925, 2, 35.
———. "Chins and Cheeks." *Popular*, 4 April 1925, 304.
———. *Cinema Acting as a Profession*. London: Standard Art Book, 1919.
———. "Develop Your Personality." *Sovereign*, March 1925, vii.
———. "Develop Your Personality!" *Pictures and Picturegoer*, August 1925, 65.
———. "Develop Your Personality." *Sovereign*, November 1925, v.
———. "The Expressions of Irene Rich." *Picture Show*, 24 May 1920, 7.
———. "The Expressions of Joyce Dearsley." *Picture Show*, 24 July 1920, 9.
———. "The Expressions of Monroe Salisbury." *Picture Show*, 22 May 1920, 7.
———. "The Expressions of Pauline Frederick." *Picture Show*, 2 October 1920, 7.
———. "The Expressions of Priscilla Deane." *Picture Show*, 1 May 1920, 9.

———. "An Extra Cash Prize." *Joy*, 21 March 1925, 29.
———. "A Few Photographs Chosen at Random from the Charming Faces Competition." *Joy*, 25 April 1925, 18.
———. "15 Minutes a Day Will Make You a Brilliant TALKER." *Sovereign*, March 1921, v.
———. "First Prize Winners in the Charm Contest." *Joy*, 23 May 1925, 2.
———. "The Flapper's Favourite." *Picturegoer*, March 1925, 34.
———. "Have You Got That £100 Look?" *Sunday News*, 26 September 1926, 7.
———. "How I Improved My Memory in One Evening!" *Sovereign*, November 1919, 81.
———. "How to Master the Piano in Three Months." *Sovereign*, January 1921, ii.
———. "How to Size People Up from Their Looks." *John Bull*, 31 May 1919, 15.
———. "Ivor Novello's Great Offer to British Girls." *Lancashire Daily Post*, 10 February 1925, 3.
———. "Ivor Novello's Great Offer to British Girls." *Daily Mail*, 11 February 1925, 5.
———. "Learn to Act for the Cinema in Your Own Home." *John Bull*, 25 October 1919, 15.
———. "Lois Laughs at Men." *Picturegoer*, April 1925, 23.
———. "A Magnetic Movie Star." *Pictures and Picturegoer*, March 1924, 30.
———. *Notable Personalities*. London: Whitehall, 1927.
———. "£100 a Week for a Face." *Sunday News*, 5 September 1926, 4.
———. "£150 and 300 Sets of Lingerie Must Be Won." *Joy*, 14 February 1925, 36.
———. "£100 Hidden in Lines on Man's Face." *Sunday News*, 22 August 1926, 4.
———. "The Secret of Being a Convincing Talker." *John Bull*, 21 June 1919, 15.
———. "The Secret of Being a Convincing Talker." *Sovereign*, June 1925, x.
———. "The Secret of Making People Like You." *Sovereign*, November 1919, 90.
———. "Seek Your Self!" *John Bull*, 14 February 1925, 9.
———. "Sir John Foster Fraser's Appeal." *Weekly Dispatch*, 29 January 1928, 9.
———. "There's Fortune in Faces for You!" *Sunday News*, 8 August 1926, 3.
———. "There's £100 for You in This Page." *Sunday News*, 15 August 1926, 3.
———. "Three Hundred Prize Winners." *Joy*, 30 May 1925, 2, 34–35.
———. "What the Mouth and Lips Denote." *Popular*, 28 March 1925, 264.
———. "Woman's Face Worth £100 to You." *Sunday News*, 29 August 1926, 4.
———. "A Woman's Woman, by de Sola." *Picturegoer*, August 1924.
———. "Your Character from Photograph." *Popular*, 14 February 1925, 43.
———. "Your Character from Photograph II.—Noses." *Popular*, 21 February 1925, 87.
———. "Your Speech Determines Your Money Value." *Sovereign*, July 1921.
———. "Your Unsuspected Self." *John Bull*, 20 September 1919, 9.
Bailey, Peter. "White Collars, Gray Lives? The Lower Middle Class Revisited." *Journal of British Studies* 38, no. 3 (1999): 273–90.
Bailey, Suzanne. "Francis Galton's Face Project: Morphing the Victorian Human." *Photography and Culture* 5, no. 2 (2012): 189–214.
Battistella, Edwin. *Do You Make These Mistakes in English? The Story of Sherwin Cody's Famous Language School*. Oxford: Oxford University Press, 2009.
Bell, Charles. *Essays on the Anatomy and Philosophy of Expression*. London: John Murray, 1824.
Bertenshaw, Mary. *Sunrise to Sunset: An Autobiography*. Manchester, UK: Printwise, 1991.
Blackford, Katherine. *Judging Character at Sight*. London: Standard Art Book, 1919.
Blayney, Steffan. "Health and Efficiency: Fatigue, the Science of Work, and the Working Body, c. 1870–1939." PhD diss., University of London, 2018.
Bogacz, Ted. "War Neurosis and Cultural Change in England, 1914–22: The

Work of the War Office Committee of Enquiry into 'Shell-Shock.'" *Journal of Contemporary History* 24, no. 2 (1989): 227–56.

Bronfman, Alejandra. "The Allure of Technology: Photographs, Statistics and the Elusive Female Criminal in 1930s Cuba." *Gender and History* 19, no. 1 (2007): 60–77.

Brown, Callum. *Religion and Society in Twentieth Century Britain*. London: Routledge, 2006.

Chase, Malcolm. "'An Overpowering Itch for Writing': R.K. Philp, John Denman, and the Culture of Self Improvement." *English Historical Review* 133, no. 463 (2018): 351–82.

Collier, John, and Ian Lang. *Just the Other Day: An Informal History of Great Britain Since the Great War*. London: Hamish Hamilton, 1930.

Collini, Stefan. "The Idea of Character: Private Habits and Public Virtues." In *Public Moralists: Political Thought and Intellectual Life in Britain, 1850–1930*, 91–118. Oxford, UK: Clarendon, 1991.

Cryle, Peter, and Elizabeth Stephens. *Normality: A Critical Genealogy*. Chicago: University of Chicago Press, 2017.

Darwin, Charles. *The Expression of Emotion in Man and Animals*. London: John Murray, 1872.

Davies, William. *The Happiness Industry: How the Government and Big Business Sold Us Well-Being*. London: Verso, 2015.

D'Cruze, Shani. "'Dad's Back': Mapping Masculinities, Moralities and the Law in the Novels of Margery Allingham." *Cultural and Social History* 1, no. 3 (2004): 256–79.

Deslandes, Paul. *The Culture of Male Beauty in Britain: From the First Photographs to David Beckham*. Chicago: University of Chicago Press, forthcoming 2021.

de Sola, Vincent. *A Thousand Dreams and Their Meanings*. New York: Mayflower, 1924.

Graves, Robert, and Alan Hodge. *The Long Week-End: A Social History of Great Britain*. London: Reader's Union, 1931.

Green-Lewis, Jennifer. *Framing the Victorians: Photography and the Culture of Realism*. Ithaca: Cornell University Press, 1996.

Hilliard, Christopher. *To Exercise Our Talents: The Democratization of Writing in Britain*. Cambridge, MA: Harvard University Press, 2006.

Laite, Julia. *Common Prostitutes and Ordinary Citizens: Commercial Sex in London, 1885–1960*. Basingstoke, UK: Palgrave Macmillan, 2012.

Law, Frederick Houk. *Mastery of Speech: A Course in Eight Parts on General Speech, Business Talking, and Public Speaking*. London: Standard Art Book, 1920.

Lears, Jackson. *Fables of Abundance: A Cultural History of Advertising in America*. New York: Basic Books, 1994.

Livesey, Ruth. "Reading for Character: Women Social Reformers and Narratives of the Urban Poor in Late Victorian and Edwardian London." *Journal of Victorian Culture* 9, no. 1 (2004): 43–67.

Montague, Nell St. John. *Revelations of a Society Clairvoyante*. London: Thornton Butterworth, 1926.

O'Rourke, Chris. *Acting for the Silent Screen: Film Actors and Aspiration Between the Wars*. London: I. B. Tauris, 2017.

Peiss, Kathy. *Hope in a Jar: The Making of America's Beauty Culture*. Philadelphia: University of Pennsylvania Press, 2011.

Price, Katy. *Loving Faster than Light: Romance and Readers in Einstein's Universe*. Chicago: University of Chicago Press, 2012.

Roper, Michael. "Between Manliness and Masculinity: The 'War Generation' and the Psychology of Fear in Britain, 1914–1950." *Journal of British Studies* 44, no. 2 (2005): 343–62.

Saler, Michael. "'Clap If You Believe in Sherlock Holmes': Mass Culture and

Stead, Lisa. *Off to the Pictures: Cinema-Going, Women's Writing, and Movie Culture in Interwar Britain*. Edinburgh: Edinburgh University Press, 2016.

Stewart, Victoria. *Crime Writing in Interwar Britain: Fact and Fiction in the Golden Age*. Cambridge: Cambridge University Press, 2017.

Thomson, Mathew. *Psychological Subjects: Identity, Culture, and Health in Twentieth Century Britain*. Oxford: Oxford University Press, 2006.

Travis, Trysh. *The Language of the Heart: A Cultural History of the Recovery Movement from Alcoholics Anonymous to Oprah Winfrey*. Chapel Hill: University of North Carolina Press, 2013.

Vernon, James. *Distant Strangers: How Britain Became Modern*. Berkeley: University of California Press, 2014.

Watts, Steven. *Self-Help Messiah: Dale Carnegie and Success in Modern America*. New York: Other Press, 2013.

Widdows, Heather. *Perfect Me: Beauty as an Ethical Ideal*. Princeton: Princeton University Press, 2018.

Woolf, Virginia. *Monday or Tuesday*. London: Hogarth Press, 1921.

[entry continued from previous page:] the Re-enchantment of Modernity, c. 1890–1940." *Historical Journal* 46, no. 3 (2003): 599–622.

2.
Identity Noir

James Purdon

But there's a question of identity involved here.
—Raymond Chandler, *The Little Sister* (1949)

In the world of hard-boiled fiction and film noir, identification is a weapon. The noir tradition is a record of misplaced, stolen, and abandoned identities, in which the verificatory processes established by modern institutions and bureaucracies for keeping track of persons and objects become the instruments both of dogged investigators, whose task is to trace and establish the identity of their quarry, and of ingenious criminals, whose task is to evade pursuit. Investigators and criminals alike depend on their ability to manipulate the media of identification. Witness Dashiell Hammett's Continental Op, who in *Red Harvest* (1929) carries a card case with a selection of false "credentials" for information-gathering purposes: "The red card was the one I wanted. It identified me as Henry F. Neill, A. B. seaman, member in good standing of the Industrial Workers of the World. There wasn't a word of truth in it."[1]

The form of noir crime fiction practiced by Hammett and his successors flourished in the interwar years, a time in American history when significant changes were beginning to reshape practices of identification in relation both to civil administration and to criminal investigation. At the beginning of the twentieth century, the identification of persons in the United States remained a highly decentralized practice, undertaken at the level of individual states. In 1903, the US Congress passed a resolution calling for the registration of births and deaths, but it would be another three decades before such a system was instituted nationwide.[2] A National Bureau of Criminal Identification had been established as early as 1896, however, its records becoming part of the foundational archive of the new Bureau of Investigation (later the Federal Bureau of Investigation) in 1908, but law enforcement still depended to a large extent on state-level organizations.[3] When the Los Angeles–based private eye Philip Marlowe wants information about a missing woman in Raymond Chandler's *The Big Sleep* (1939), he impersonates an officer not of the FBI but of the "Police Identification Bureau." (California's State Bureau of Criminal Identification and Investigation, founded in 1905 and closed four years later due to lack of funding, had reopened in 1917 on a firmer footing.)[4]

Problems of true and false identification were central to noir because they were increasingly central to the society that produced it. Whether understood as a genre, an aesthetic, or a tradition, noir became established as the dominant form of American crime writing just as new institutions and procedures for verifying, codifying, and recording identity were becoming commonplace in American society. Noir representations of identity problems are of particular interest within the cultural history of identification because they record and dramatize the implications of several concurrent shifts in the construction of identity as a matter of public record. The fundamental question of what determines an individual's identity plays out in these works of fiction as a struggle between the various claims of psychological interiority, social recognition, official bureaucracy, and mass media. The rise of personal identification as an object of bureaucratic attention in the institutions of the early twentieth-century state and in the files of commercial organizations helped to shift the responsibility for establishing and maintaining the link between bodies and their histories decisively away from narrative practices, such as social reputation and autobiography, and toward archival ones, such as identity cards, social security numbers, driver's licenses, passports, and other ubiquitous "papers." The old autograph signature, always highly vulnerable to forgery, was supplemented by a panoply of new probative methods such as photography, anthropometry, and fingerprinting, each of which claimed a superior ability to establish the identity (that is, the sameness) of a body over time. As these new methods came to be accepted as essential tools of state administration, they also began to impinge on the autonomy and agency of the self as an embodied subjectivity. The more an officially recognized identity became a matter of documentation and data, the less power narrative explanation had to render an account of the self that might challenge this authoritative official construct. The ascendancy of bureaucratic identification enables not only the police inquiry and the novel of criminal detection but the frame-up and the wrong-man story as well.

Noir fiction and film played a leading role in the imaginative project of interpretation and critique that enabled twentieth-century readers and cinema audiences to understand the capabilities and shortcomings of these new systems of identity control. The writers and filmmakers who shaped noir's narrative conventions and its visual styles took the practice of identification seriously, in accordance with the new sophistication of technical means for identifying persons and the newly powerful role those techniques played in crime and the detection of crime. In the seminal essay "The Simple Art of Murder," it is to a question of identification that Chandler turns when he wishes to draw a line between the Golden Age of English country-house detective fiction and the grittier, more realistic productions of the new American hard-boiled type. Disparaging A. A. Milne's *The Red House Mystery* (1922), in which an English gentleman is murdered by his secretary after playing a trick on his friends by impersonating his long-lost brother, Chandler complains that "no competent legal identification is offered" to establish the identity of the corpse: "Identification is a condition precedent to an inquest. It is a matter of law. Even in death a man has a right to his own identity."[5] To Chandler, writing in the mid-1940s, Milne seemed representative of an earlier generation of writers who had not paid enough attention in their drawing-room mysteries to the procedures of identification that form the basis of modern police work.

"Even in death a man has a right to his own identity." Yet as Chandler's emphasis on legal identification suggests, that right is not unlimited. In noir, one has a right to an identity only insofar as it can be corroborated by agents of the law; this is why early noir fiction, like the nineteenth-century detective novels that laid the foundations of the genre, invariably ends up affirming the principle of identifiability. The murderous show girl may start a new life as a banker's elegant wife; the missing homicide suspect may turn up with a new face from a Mexican clinic; but an altered name or a bout of plastic surgery is never enough to fool a Continental Op or a Marlowe. Finding people who would prefer not to be found is, after all, among the private eye's characteristic functions. He, if no one else, confirms that identities are, in practice as well as in theory, fixed and traceable. The interest shown in identification by Hammett and Chandler is procedural rather than existential. If their stories frequently imply the fallibility of modern identification practices, their stoic investigators seldom have the time or the inclination to reflect on how those practices transform the relationship between the private citizen and society or how they affect the citizen's understanding of his or her own identity. Such reflection requires a kind of distance unavailable to the private eye, who may be exceptional within his society but is nonetheless shaped by and embedded in it. Marlowe's occasional reflections on the social conditions of modernity do not extend as far as a critique of the official administration of identity, since his own role as a private eye both depends on and reinforces the authority of identification. A Marlowe who stopped to engage in existential theorizing about the identity regime in which he operates would not be Marlowe at all.

By the beginning of the 1940s, however, the best of the new noir fiction had started to take an interest in identity as a problem of more than procedural significance. If classic hard-boiled fiction presented an investigator struggling to unravel a plot from the outside, the protagonists of the psychological noir of the 1940s were more frequently faced with the difficulty of understanding a plot in which they find themselves unwillingly implicated. The protagonist of Cornell Woolrich's *The Black Curtain* (1941), for instance, recovers consciousness on a crowded street, having been knocked out by a fragment of falling masonry. His name, he knows, is Frank Townsend—but the silver cigarette case in his pocket is unfamiliar, and the initials inside his hat read "D. N." He makes his way home to find his apartment empty and his wife absent. In tracking her down, he discovers that he has been missing for more than three years. With no recollection of his immediate past, Townsend goes back to work but soon finds himself in the familiar noir predicament of being hunted by a shadowy, gun-toting pursuer.[6]

The amnesiac Townsend is faced with a problem of identity unlike any encountered by Sam Spade or Philip Marlowe. His task is not to trace and identify another person but rather to recover the continuity of his own consciousness by establishing the identity, the sameness, of his present self and the unremembered "D. N." whom he has embodied for the last three years. And he knows that meanwhile his mysterious pursuers will be working to discover the new identity that he appears to them to have adopted. When Townsend tries to shake off his pursuer by quitting his job, he quickly realizes that his "address . . . , his name, and all other pertinent information about him were on file down where he had worked, accessible to the most casual inquiry" (23). Townsend, for his own part, has no recourse to such documentation; the revelation of his former identity comes by chance after he takes up residence on the street where his accident occurred and is eventually noticed by his former girlfriend. Townsend thus discovers that he has been living as Dan Nearing, a drifter now wanted for the murder of his former employer on a private estate in the country: "Now the curtain had lifted and he saw what lay before him. No personal vengeance, no private enemy stalking him from out of the miasmas of the past. That had been organized society itself. That man must have been of the police" (79). In order to recover his identity Townsend revisits the scene of the crime, where he discovers that he has been framed by the man's heirs, and manages—at great personal risk—to clear his name.

While Townsend's police pursuers are empowered to trace him using the systems of "organized society," linking names, faces, and actions by means of discrete data kept "on file," Townsend's own approach to reestablishing his identity relies on a process akin to the psychoanalyst's abreaction, in which repressed memories are recuperated into conscious awareness by being reintegrated into the subject's continuous personal narrative. Seeking to establish this continuity by returning to the estate of the murdered man, Townsend comes to understand his task precisely as a form of storytelling: "This was the story's end, one way or another. This was the night. This was the time. This was the place.... He thought of the strange story that he'd lived, his own story.... Whatever happened, he'd never be quite like other people again" (120).

Throughout the novel Townsend is driven by his inability to reconcile the public record of his apparent crime with his own private narrative. The story simply doesn't fit. His frustration at the apparently unresolvable difference comes to a head at the end of the book's middle section, when he resolves to return to the scene of the murder:

> I know I didn't do that. Don't ask me to back that up. I can't. I know three people saw me. I know it's in the papers, and on the cops' blotters. I don't care. I don't care if the whole world says I killed that man. *I* say I didn't. The *me* that's in me says I didn't! I won't lie down and let them tell me different. (80)

What Townsend insists on is a right to identity that goes further than that found in Chandler's procedural tying up of loose ends. His claim to his own identity rests not on being properly matched to a set of documents but on his prerogative to give an alternative account of himself that conflicts with the official records held on him by the police department and the public record printed in the newspaper. In its self-conscious exploration of the disjuncture between informational and narrative definitions of identity—the one understood as the objective product of a set of records at a given moment in time, the other as the outcome of an ongoing subjective awareness of self-conscious continuity—Woolrich's form of noir recognizes that the modern regime of identification is fallible. But it also goes further. Townsend's indignation at the apparently incontestable story told by the papers and the cops' blotters shows how the imbalance of power between the objective claims of identification and the subjective experience of identity manifests at the level of individual psychology as a division of the self: "the *me* that's in me." Far-fetched as Townsend's predicament is, it is a good example of a shift in noir thinking about identity. By revealing the yawning gap between the private experience of selfhood and the uncontrollable proliferation of the self's public identities, Townsend's dilemma shows how highly organized systems of identification began to register in fiction not merely as a matter of criminological procedure but as the symbol of a crisis in the modern psyche.

One term that has frequently been used to describe the cultural expression of such early twentieth-century crises of the psyche is *modernism*. Although literary scholars have usually been inclined to retain some sort of distinction between the high-modernist productions of the European and American avant-gardes of the early twentieth century and the widely consumed products of popular writers in those regions, the past twenty years or so have seen the emergence of a more expansive scholarly interest in the porosity of these categorical boundaries and in the currents of influence that traversed them. Noir fiction seems an apt test case for such inquiries, since, as Lee Horsley has suggested, while noir generally eschewed the "aesthetic sophistication and deliberate difficulty" of pre-1930s modernist avant-gardes, "modernist techniques as well as themes helped to shape literary noir, encouraging, for example, the use of irony, non-linear plots, subjective narration and multiple viewpoints."[7] Drawing formally on both realist and modernist narrative techniques and fluctuating between propulsive dime-novel plotting and existential reflection, noir fiction constitutes one of those forms of "secondhand" modernism that Paula Rabinowitz has identified as central to the consolidation of pulp fiction in 1930s America, offering readers the opportunity to enjoy (or writers the opportunity to surreptitiously slip those readers) some modernist innovation under the cover of gritty realism.[8]

In this respect, noir's obsession with identification might be said to have materialized the modernist identity crisis, transforming it from an abstract problem of literary aesthetics—how can the experience of a continuous consciousness best be represented in writing?—into a concrete problem of everyday experience: what happens when the right to an identity is determined less by the experience of continuous consciousness than by official writing? Not everyone, to be sure, suffered from identity-rupturing amnesia, but a great many people were beginning to understand how the new power attributed to identification as a bureaucratic and technical practice had irrevocably altered the relationship between citizen and state. Craig Robertson has written compellingly about the "passport nuisance" of the 1920s, when increasing numbers of American citizens traveling abroad in the aftermath of World War I found themselves required to justify their right to an identity. Newspapers carried frequent outbursts of frustration from individuals who regarded as a personal affront both the bureaucratic nature of the application process and the generalized attitude of suspicion it seemed to imply: "The humiliation stemmed from the request for . . . documents over and above an individual's word. Travelers took this rejection of their word (and appearance) as a sign that officials considered them dishonest and untrustworthy—a response grounded in the association of identification documents with suspect individuals such as criminals."[9] If the passport remained a voluntary credential in the sense that it was required only by those undertaking foreign travel, the following decade saw the introduction of a compulsory credential in the form of the social security number. Mandated as part of Roosevelt's New Deal Social Security Act in 1935, the social security number added to a growing list of documents—draft cards, driver's licenses, automobile registration papers, insurance certificates—that had already contributed to what Robertson calls "the general paperization of everyday life."

One poem published that year expresses the common anxiety that the proliferation of means for capturing and storing identifying information about individuals might result in the erosion of something essential to individuality itself. Kenneth Fearing's noir-shadowed "Escape" makes explicit the connection between the ascendancy of identification and the loss of personal identity that many of his fellow citizens, in decrying the "passport nuisance," had already intuited:

> Acid for the whorls of the fingertips; for the face, a
> > surgeon's knife; oblivion to the name;
> > eyes, hands, color of hair, condition of teeth, habits,
> > > haunts, the subject's health;
> > wanted or not, guilty or not guilty, dead or alive, did
> > > you see this man

> Walk in a certain distinctive way through the public streets
> or the best hotels,
> turn and go,
> escape from collectors, salesmen, process-servers, thugs; from
> the landlord's voice or a shake of the head; leave
> an afternoon beer; go from an evening cigar in a
> wellknown scene,
> walk, run, slip from the earth into less than air?
>
> . . . .
>
> But something must be saved from the rise and fall of the
> copper's club; something must be kept from the
> auctioneer's hammer; something must be guarded
> from the rats and the fire on the city dump;
>
> . . . .
>
> And in what
> disguise did the soiled, fingerprinted, bruised,
> secondhand,
> worn-down, scarred, familiar disguise
> escape?
>
> No name, any name, nowhere, nothing, no one, none.[10]

Fearing's ironic title plays on noir's capacity to offer the mystery story's escapist pleasures while laying claim to the legitimacy of realism—what Chandler, in "The Simple Art of Murder," calls "the authentic flavor of life as it is lived."[11] Yet, like much post-hard-boiled noir, the poem is resolutely skeptical about the claim to authenticity. There is no hint here that the inauthenticity of modern life might be redeemed by the discovery of a stable, authentic self. Instead, there is only the troubled thought that the modern self might prove to be nothing more than the limited set of marks, habits, and conventions that serve as identifying features. The question remains: what or who will be left if the would-be escapee manages improbably to cast off those distinguishing characteristics? "Escape" channels the anxiety that, for the individual entangled in the identificatory nets of modern society, there may be no authentic self that matters outside the data of the police dossier, the clichés of hard-boiled fiction, and the standardized masculine rituals of the afternoon beer and the evening cigar. With each layer of identifying features it strips away, the poem's insistence that "something must be saved . . . ; something must be kept" breaks down until, in the end, the "disguise" of identity falls away only to reveal another "disguise" alongside the recognition that the only true escape now possible is the self's total negation, its transformation into something "less than air," into "nothing, no one, none."

Like Woolrich—who before turning to popular crime writing had written half a dozen novels in the jazz-age idiom of F. Scott Fitzgerald—Fearing was by inclination a modernist who eventually found his literary niche in noir. Fearing, however, framed his literary work in explicitly political terms. As an associate editor of the communist-aligned *Partisan Review*, he had combined an interest in modernist experimentation with a strong commitment to revolutionary politics, helping to publish work by John Dos Passos, Richard Wright, Josephine Herbst, and Muriel Rukeyser as well as André Gide, Louis Aragon, György Lukacs, and Nikolai Bukharin.[12] His own poetry used a loose free verse, modeled on the Whitmanesque long line, to carry a witheringly satirical depiction of American consumer capitalism, frequently evoking the noir milieu that provides the setting for much of his later fiction.

With its knowing pastiche of tropes and settings from the pulps of the previous decade, Fearing's best novel, *The Big Clock* (1946), has usually been regarded as an eccentric minor entry in the canon of American crime fiction, tending to strike readers according to taste as either "an intellectual's crime novel" or a "spoof *roman noir*."[13] But to call Fearing's novel a spoof is, I think, to underestimate it. *The Big Clock* is serious parody, which condenses and clarifies noir fiction's obsession with modern problems of identity and identification while also subverting early noir's tendency to resolve those problems through the intervention of the incorruptible private eye, that key noir protagonist whose characteristic function is precisely to reassert order and justice by tracing and identifying people.

The novel's protagonist, true-crime magazine editor George Stroud, begins an illicit affair with a mysterious blonde. After their second weekend rendezvous at an upstate hotel, he drives her back to Manhattan, where from the shadow of a streetlight he watches her enter her apartment with a man he recognizes as his own employer, a wealthy and well-connected publishing magnate. A little disheartened, George goes for a drink before returning home to his wife and children. At breakfast on Monday, however, he picks up the morning paper to find a headline reporting the woman's brutal murder. In a daze, he catches the train to the office, where things take an unexpected twist. The murderous publishing magnate has formulated a plan to divert suspicion by framing the shadowy figure he noticed lurking in the street, and the editor finds himself assigned to identify and implicate the mystery man: himself.

Fearing's masterstroke, in *The Big Clock*, was to make the investigator and the suspected criminal the same man, condensing the investigation plot into a tense circular pursuit. George is faced with the task of identifying and tracing a mystery man whom he knows to be *himself*; at the same time, he must manage the investigation in such a way as to avoid either implicating himself or raising suspicion. It is this ouroborotic plot that makes *The Big Clock* the most sophisticated and sustained exploration of identity in noir fiction. *The Big Clock* is a highly self-conscious piece of work, both in the sense that it reflects in sophisticated ways on the conventions of its own genre and the media environment that sustains and satisfies the appetites of its readers and in the sense that the troubled relationship between the private self and the tokens of a public identity is the unifying theme of those reflections. What raises *The Big Clock* above mere burlesque is the mutual reinforcement in its narrative of these two forms of self-consciousness. The circumstances that force its protagonist into an awareness of the precarious relationship between identity and identification enable the novel's own reflections on the conventions of the noir tradition, while the novel's subversion of those very conventions serves to focus attention on the problematic of modern selfhood articulated by the genre of noir as a whole.

Just as George Stroud finds himself impossibly suspended between the obligation to orchestrate the hunt for his shadow and the necessity of undermining that pursuit, the novel operates as a satisfying wrong-man thriller even as it parodies the genre to which it belongs. George is hardly the incorruptible private eye envisioned by Chandler: his extramarital dalliance is what gets him into trouble in the first place, and no Sam Spade or Philip Marlowe is going to come along to get him out of that trouble. Once the financial resources of the pulp publishing corporation have been turned against him, his only recourse is to turn saboteur, delaying the progress of his investigation by preventing his own identification as the man in the shadow of the streetlight. In the end, the pressure that brings the novel to its conclusion is not the dogged pursuit of an implacable detective but the impersonal forces of commerce: the newspaper proprietor simply drops the investigation, having been forced out of his position by a corporate merger.

In *America Is Elsewhere: The Noir Tradition in the Age of Consumer Culture*, Erik Dussere proposes that the noir tradition can be understood as a manifestation of the yearning for an "authentic" national identity amid America's postwar consumer boom, a reading that reveals continuities in noir writing from hard-boiled to cyberpunk. Yet Dussere may be a little wide of the mark when he finds in *The Big Clock* an "authentic version of male identity . . . , the detective locked in a schizoid battle with the inauthentic executive on the battleground that is George Stroud."[14] George is no detective. Employed to trace himself, he spends his time in misdirection rather than investigation. It is his very inauthenticity, his lack of a distinctive, stable, incorruptible identity, that ultimately enables him to avoid being framed for the crime of murder, since his escape relies on the fact that he is sufficiently indistinguishable from any number of organization men to elude identification just long enough for the whole matter to blow over. At a tense point in the investigation, George listens on the telephone as one of his reporters passes on an eyewitness description that is as undistinguished as it is unflattering:

> He was a smug, self-satisfied, smart-alecky bastard just like ten million other rubber-stamp sub-executives. He had brown hair, brown eyes, high cheekbones, symmetrical and lean features. His face looked as though he scrubbed and shaved it five times a day. He weighed between one sixty and one sixty-five. Gray tweed suit, dark blue hat and necktie.[15]

In the first pages of the novel, George confronts his reflection in the shaving mirror. The weary recognition of time's passing—"a tuft of gray on the right temple had stolen at least another quarter-inch march"—is quickly replaced by reverie as he imagines himself as the object of his younger colleagues' attention in years to come:

> *Who's that pathetic, white-haired old guy clipping papers at the desk over there?* asked a brisk young voice. But I quickly tuned it out and picked another one: *Who's that distinguished, white-haired, scholarly gentleman going into the directors' room?* (8)

George's daydream, with its transvaluation of a "pathetic, white-haired old guy" into a "distinguished, white-haired, scholarly gentleman," neatly suggests the shortcomings of a model of identification that neglects personal narrative in favor of the outward signs of identity. One man's doddery old-timer is another man's respectable scholar. Sinking further into his vision of the future, however, George imagines himself as a figure of some notoriety: "He might have been one of the biggest men in aviation today, only something went wrong. I don't know just what, except that it was one hell of a scandal. Stroud had to go before a Grand Jury, but it was so big it had to be hushed up, and he got off. After that, though, he was through" (9).

This mirror play occurs on the morning of George's illicit rendezvous and is the first of many occasions in the novel when its protagonist will be confronted with a reflected version of himself. In the privacy of the shaving mirror, of course, none of this matters much: the more undesirable of his reflections can easily be "tuned ... out" so as to preserve some measure of dignity. Throughout the novel, however, George will be compelled to acknowledge similar third-person accounts of his identity, the most egregious of which takes the form of a profile compiled by the journalists George has dispatched on his own trail and scrawled on a chalkboard in the magazine's Manhattan office:

> I went over to the board, topped with the caption: X.
> In the column headed: "Names, Aliases," I read: George Chester?
> Under "Appearance" it said: Brown-haired, clean-cut, average height and build.
> I thought, thank you, Ed.
> "Frequents": Antique shops, Van Barth, Gil's. At one time frequented Gil's almost nightly.
> It was true, I had.
> "Background": Advertising? Newspapers? Formerly operated an upstate tavern-resort.
> Too close.
> "Habits": Collects pictures.
> "Character": Eccentric, impractical. A pronounced drunk.
> . . . .
> I said, standing beside the word-portrait of myself: "We seem to be getting somewhere."
> (116)

The director John Farrow made this chalkboard profile, or rather a version of it, a centerpiece of the set for his 1948 film adaptation of *The Big Clock*, and it is easy to guess why. Just as Farrow inserted a literal big clock in the publishing company foyer as a visual representation of the novel's metaphorical one (George's expansive term for all the systems and structures that govern modern life in the interests of commerce and capital), the chalked-up profile visualizes the set of authoritative categories—names, locations, habits—into which the modern citizen-subject is broken up for the purposes of identification. It also serves as a reminder of the jarring, uncanny, and unpleasant reaction that usually accompanies the experience of confronting such data doubles. Yet what makes George Stroud doubly interesting as a protagonist is that the forced alienation of his identity doesn't simply unsettle him; in a certain way, it dissolves his faith in identity entirely. It isn't that George feels himself to have a more legitimate claim to his authentic self than does the chalkboard but rather that the whole charade of self-pursuit leaves him wondering whether he has an authentic self at all. As George comes to realize, he is inauthentic from the ground up. Even his private family life—shared with his wife Georgette and daughter Georgia—seems designed to parody an adman's fantasy of postwar domestic harmony, with each member of the family deriving her name and identity from George himself, while his seemingly authentic masculine identity as husband and father is, as his infidelity reveals, just another performance.

By the end of the novel, George has successfully eluded the plot to frame him, but only at the cost of discovering that, despite his ingrained sense of his own distinctiveness, he appears to others as little more than a cipher—first as the impractical drunkard profiled by his investigative colleagues and later as the nondescript gray-suited company man in the eyewitness description. What else remains? A Sam Spade is a Sam Spade (as his name suggests) and a Marlowe is a Marlowe. But who is George Stroud? An answer, of a kind, comes in a seemingly trivial scene near the end of the novel. Having learned that he is in the clear, George phones home and talks to his six-year-old daughter, Georgia, who demands a story. The parable he improvises sounds nonsensical, but it bears the unmistakable imprint of George's recent experience with the tangled skeins of modern identification:

> "Hello, George, you have to tell me a story. What's her name?"
> "Claudia. And she's at least fifteen years old."

"Six."

"Sixteen."

"Six. Hello? Hello?"

"Hello. Yes, she's six. And here's what she did. One day she started to pick at a loose thread in her handkerchief, and it began to come away, and pretty soon she'd picked her handkerchief until all of it disappeared and before she knew it she was pulling away at the yarn in her sweater, and then her dress, and she kept pulling and pulling and before long she got tangled up with some hair on her head and after that she still kept pulling and pretty soon poor Claudia was just a heap of yarn lying on the floor."

"So then what did she do? Hello?"

"So then she just lay there on the floor and looked up at the chair where she'd been sitting, only of course it was empty by now. And she said, 'Where am I?'" (170)

As "Claudia" pulls away at her handkerchief, she finds that the stuff of her own identity is more entangled with the surrounding world than she has previously understood. George's story is a yarn about yarns, about how when we begin to unpick the stories we tell to ourselves about ourselves we discover that a seemingly substantial identity can turn out to be nothing more than a heap of yarns. It is a lesson George has been forced to take to heart, and one that the noir of the 1940s continually impresses on its readers. If Townsend, in Cornell Woolrich's *The Black Curtain*, knows he is innocent because of "the *me* that's in me," by the end of *The Big Clock* George Stroud is beginning to wonder ("Where am I?") whether there's any *me* in him at all. The continuity between the narrating self and the self as subject of that narration has been lost, perhaps irrevocably.

Such problems of identity took on new urgency in the noir writing of the 1940s. These new narratives told neither the old realist or naturalist story about characters whose true selves are gradually revealed in their public dealings and intimate connections nor the new modernist story about provisional subjectivities formed moment by moment in the sensory whirl of the twentieth century. Catalyzed by new technological and administrative forms of identity description, noir instead began to tell stories about the precariousness of narrative itself in a world designed to keep personal identities straight by technological means. Whereas having an identity had once meant being able to give a continuous account of a developing self, hereafter it would mean being able to match one's body, speech, and behavior to a set of documents and data points held on file by an external authority.

That transformation in the way identity was conceived set the scene for the shifting, unstable protagonists of the following decade's psychological thrillers, such as the sociopathic sheriff's deputy Lou Ford in Jim Thompson's *The Killer Inside Me* (1952) and the manipulative, ruthless Tom Ripley in Patricia Highsmith's *The Talented Mr. Ripley* (1955). When the protagonist of Highsmith's *The Blunderer* (1954) is described as feeling "like nothing but a pair of eyes without an identity behind them," his predicament can be understood as the logical consequence of the hollowing out of identity by identification that had been charted in Woolrich's and Fearing's fiction.[16] For Highsmith's protagonists, the fantasy of escaping from an established identity always goes hand in hand with the disinhibition of murderous urges, but the man who cheats his way out of his own identity is dangerous not just because of any crimes he might commit against individuals. Rather, he is dangerous because his transgression of the legal and social systems that establish and guarantee the right to an identity exposes the weaknesses in those systems. By revealing the flawed processes through which the fiction of stable identity is constructed and maintained, he undermines the whole social order that it had been the business of the Chandler private eye to fortify. If, in Chandler's famous formulation, the mean streets of the modern city could be made safe only by a man who was not himself mean, the inheritors of the noir tradition understood that the greatest threat to that safety was a man who was not himself at all.

Notes

1. Dashiell Hammett, *Red Harvest*, in *The Complete Novels* (New York: Library of America, 1999), 8.
2. David I. Kertzer and Peter Laslett, eds., *Aging in the Past: Demography, Society, and Old Age* (Berkeley: University of California Press, 1995), 331.
3. Jonathan Finn, *Capturing the Criminal Image: From Mug Shot to Surveillance Society* (Minneapolis: University of Minnesota Press, 2009), 37.
4. David Q. Burd, "The Laboratory Section of the California State Bureau of Criminal Identification and Investigation," *Journal of Criminal Law, Criminology, and Police Science* 43, no. 6 (March–April 1953): 829–33.
5. Raymond Chandler, "The Simple Art of Murder," *Atlantic Monthly* 174, no. 6 (December 1944): 53–59.
6. Cornell Woolrich, *The Black Curtain* (1941; New York: Ballantine Books, 1982).

Further references to this edition are given in parentheses in the text.

7. Lee Horsley, *The Noir Thriller* (Basingstoke, UK: Palgrave Macmillan, 2001), 3.

8. Paula Rabinowitz, *American Pulp: How Paperbacks Brought Modernism to Main Street* (Princeton: Princeton University Press, 2014), 38.

9. Craig Robertson, *The Passport in America: The History of a Document* (New York: Oxford University Press, 2010), 216.

10. Kenneth Fearing, *Poems* (New York: Dynamo, 1936), 47–48.

11. Chandler, "Simple Art of Murder," 53–59.

12. Although he has sometimes been identified as a founding editor of the magazine, Fearing is first listed as editor in the notes on contributors in the January–February 1935 issue (vol. 2, no. 6).

13. Richard Elman, "Durable Time-Piece," *The Nation* 231 (September 20, 1980): 256; Rabinowitz, *American Pulp*, 152n36.

14. Erik Dussere, *America Is Elsewhere: The Noir Tradition in the Age of Consumer Culture* (New York: Oxford University Press, 2014), 93.

15. Kenneth Fearing, *The Big Clock* (1946; New York: New York Review Books, 2006), 122. Further references to this edition are given in parentheses in the text.

16. Patricia Highsmith, *The Blunderer* (1954; New York: W. W. Norton, 2001), 223.

References

Burd, David Q. "The Laboratory Section of the California State Bureau of Criminal Identification and Investigation." *Journal of Criminal Law, Criminology, and Police Science* 43, no. 6 (March–April 1953).

Chandler, Raymond. "The Simple Art of Murder." *Atlantic Monthly* 174, no. 6 (December 1944).

Dussere, Erik. *America Is Elsewhere: The Noir Tradition in the Age of Consumer Culture*. New York: Oxford University Press, 2014.

Elman, Richard. "Durable Time-Piece." *The Nation* 231 (September 20, 1980).

Fearing, Kenneth. *The Big Clock*. New York: New York Review Books, 2006. Originally published 1946.

———. *Poems*. New York: Dynamo, 1936.

Finn, Jonathan. *Capturing the Criminal Image: From Mug Shot to Surveillance Society*. Minneapolis: University of Minnesota Press, 2009.

Hammett, Dashiell. *Red Harvest*, in *The Complete Novels*. New York: Library of America, 1999.

Highsmith, Patricia. *The Blunderer*. New York: W. W. Norton, 2001. Originally published 1954.

Horsley, Lee. *The Noir Thriller*. Basingstoke, UK: Palgrave Macmillan, 2001.

Kertzer, David I., and Peter Laslett, eds. *Aging in the Past: Demography, Society, and Old Age*. Berkeley: University of California Press, 1995.

Rabinowitz, Paula. *American Pulp: How Paperbacks Brought Modernism to Main Street*. Princeton: Princeton University Press, 2014.

Robertson, Craig. *The Passport in America: The History of a Document*. New York: Oxford University Press, 2010.

Woolrich, Cornell. *The Black Curtain*. New York: Ballantine Books, 1982. Originally published 1941.

3.
"The Ghosts of Individual Peculiarities"
Murder and Interpretation in Dickens

Andrew Mangham

On 9 October 1857 two lightermen working on the Thames were rowing their boat under Waterloo Bridge when they saw a carpetbag resting on one of the structure's buttresses. They pulled up the bag "as soon as possible in order to seize what they considered a prize."[1] It was very heavy, and, on arriving home, they opened it: "A horrible spectacle presented itself to their view—portions of a human body, bones from which the flesh had been rudely torn, and garments saturated with blood."[2] According to *The Times*, the body had evidently belonged to a man and had been made "a thing of shreds and patches"; a preliminary examination of his pieces of clothing suggested that he had moved "in the upper class of society."[3] The lightermen took the bag and its grisly contents to Bow Street station, and the remains were subsequently examined by division surgeon Mr. Paynter:

> The result of the examination . . . showed the bag to contain a great number of the different portions of a human body. . . . The parts found consisted of the legs, arms, nearly the whole of the spinal column, the buttock joints, and the shoulder joints. The whole of the head and several cervicals of the vertebrae, the hands, and the feet were absent. With regard to the condition of the remains, it was found that the greater portion of the flesh had been very roughly removed. . . . From the absence of the head it is impossible to guess even at the age of the unfortunate man, but from the appearance of the bones of the limbs Mr. Paynter is of the opinion that the deceased was a full-sized robust man.[4]

The remains were sent to Alfred Swaine Taylor, professor of medical jurisprudence at Guy's Hospital. Taylor was the author of several books on medical jurisprudence, including *Elements of Medical Jurisprudence* (1836), *Medical Jurisprudence* (1845), and *On Poisons* (1848). Although he was cautious about the exact cause of death, Taylor ascertained,

> in one portion of the left side of the chest . . . , there is an aperture in the flesh presenting the appearance of a stab. . . . Assuming that this wound would have been inflicted during life, it would have penetrated the heart, and have produced rapid, if not immediate, death. The muscles of the chest through which this stab had passed were for some space around of a dark red colour, evidently produced by blood which had been effused as a result of this wound. This . . . led me to the conclusion that this wound was inflicted on the deceased either during life or within a few minutes of death. . . . As the organs of the chest and abdomen are not forthcoming, any opinion on the cause of death must be a matter of speculation.[5]

A rumor got abroad that the remains might have come from a cadaver belonging to an anatomy school and that a group of medical students had placed the remains on Waterloo Bridge as a prank. But Taylor insisted that "the portions of the 23 pieces of the body presented no appearance of having undergone dissection for the purposes of anatomy. . . . The clearest examination, coupled with the knowledge derived from an experience of seven years spent in the study of anatomy by dissections, leads me to the conclusion that these remains have not been employed for any anatomical purpose whatever."[6] Taylor did find evidence that the body had been kept in common salt, probably in order to slow the process of putrefaction while the body was stored; the corpse had also been boiled, possibly in order to remove the flesh from the bones and thus make it difficult to identify the body. If such was the intention, Taylor said, it was successfully implemented. He was unable to determine how long the deceased had been dead. "On this point," he said, "only a speculative opinion can be given. . . . Those changes in the animal matter on which we are accustomed to rely for evidence of the period of death have been suspended. Still, an examination of the deep-seated parts of the flesh . . . has led me to the conclusion that the person of whose body these remains were a part may have been dead three or four weeks prior to that date."[7] At the inquest, the coroner praised Taylor for having so completely anticipated every question that could arise. The jury returned a verdict of willful murder by some person or persons unknown. *The Times* was optimistic that "notwithstanding the great care and the evident attempt to conceal the fact which the perpetrators of this diabolical deed had taken, murder will out."⁸ But the murder never did out. Not only were the perpetrators never found, but the unfortunate man in the carpetbag was never identified.

Regardless of the shortcomings of the Waterloo Bridge investigation, Taylor's science demonstrates how forensic medicine had developed by this time into a sophisticated concern with telling stories—with excavating evidence, piecing clues together, and reading the past. It is the argument of this chapter that as such, medical jurisprudence had much in common with the period's literature, especially that of Charles Dickens, whose urban narratives interrogated the meanings of everyday objects and of human behavior as well as the influences of past actions. In an 1860 article published as part of *The Uncommercial Traveller* series (1860–68), the author mentioned the man whose remains had been discovered at Waterloo Bridge—the "chopped up murdered man," as Dickens called him.⁹ "Night Walks" is an example of the way Dickens often used his noctambulations as inspiration for literary explorations of London's more dubious places and inhabitants. He explained that his usual practice when afflicted with insomnia involved "getting up directly after lying down" and then "going out, and coming home tired at sunrise."¹⁰ He describes walking across Waterloo Bridge on one of these outings:

> There was need of encouragement on the threshold of the bridge, for the bridge was dreary. The chopped up murdered man, had not been lowered with a rope over the parapet when those nights were; he was alive, and slept then quietly enough most likely, and undisturbed by any dream of where he was to come. But the river had an awful look, the building on the banks were muffled in black shrouds, and the reflected lights seemed to originate deep in the water, as if the spectres of suicides were holding them to show where they went down. The wild moon and clouds were as restless as an evil conscience in a tumbled bed, and the very shadow of the immensity of London seemed to lie oppressively upon the river.¹¹

Although there is evidence that some of Taylor's forensic methods are mirrored in Dickens's dark prose, there are also some significant reconfigurations of the professor's science. Of course, major dissimilarities stem from the fact that Taylor and Dickens had entirely different objectives. Even in his journalism, Dickens was fairly liberal with his use of symbolism and metaphor, while Taylor's solemn responsibilities as a forensic examiner appear to have precluded any mode of interpretation that did not adhere to empirical evidence.

That the two had different aims in their enterprises, however, should not distract us from the explorative work that Dickens's writing performs with a set of tools similar to that used by Taylor. Reading the past, telling stories, and interpreting both actions and identities all belong to a complex exploration of how each of these processes works. For example, the idea that the specters of suicides hold lights to signal where they went down is an extraordinary image that embodies the forensic belief that actions leave evidentiary trails. The fact that Dickens repurposes the idea as a ghostly simile typifies a belief he has in such traces as having the fictive quality of existing beyond the currents of empiricist observation. As John Bowen notes, "Dickens's fiction is fascinated by what is dead but will not lie down, in things or people or people-things who cross or trouble the boundaries between what was, what is, and what may be living."[12] George Augustus Sala once said that Dickens "liked to talk about . . . the latest murder and the newest thing in ghosts."[13] The description of Waterloo Bridge in "Night Walks" indicates that the author did not see these two fascinations as unrelated: he perceived murders to leave traces that were ghostly in the way they haunted the margins of interpretation. Sala's ironic idea of a "newest thing in ghosts" is reflected in Dickens's Thames, where the lights of the modern city provide an index to the sad stories of past suicides. Paradoxically, such acts of self-murder are both part of the modern world and untouched by its progress; their existence is evidenced through a simile that relies on the language of forensic enlightenment, yet in making the image a supernatural one, Dickens renders the act of interpretation aware of its own construction.

To better understand the trace-like nature of murder in Dickens's narratives, it is worth exploring the more conceptual intersections that his works create between history and crime. In *The Mystery of Edwin Drood* (1870), the unfinished murder story begins with a question of identification that is answered by the past:

> An ancient English Cathedral Town? How can the ancient English Cathedral town be here! . . . For sufficient reasons which this narrative will itself unfold as it advances, a fictitious name must be bestowed upon the old Cathedral town. Let it stand in these pages as Cloisterham. It was once possibly known to the Druids by another name, and certainly to the Romans by another, and to the Saxons by another, and to the Normans by another; and a name more or less in the course of many centuries can be of little moment to its dusty chronicles.[14]

Cloisterham is in fact better identified through traces it retains of its long-dead inhabitants:

> In a word, a city of another and a bygone time is Cloisterham, with its hoarse Cathedral-bell, its hoarse rooks hovering about the Cathedral tower, its hoarser and less distinct rooks in the stalls far beneath. Fragments of old wall, saint's chapel, chapter-house, convent and monastery, have got incongruously or obstructively built into many of its houses and gardens, much as kindred jumbled notions have become incorporated into many of its citizens' minds. All things in it are of the past.[15]

As in other late novels such as *Bleak House* (1852–53), *Little Dorrit* (1855–57), and *Great Expectations* (1860–61), in *Edwin Drood* the past becomes an oppressive weight on the present. The text creates a fictional world in which the manners and customs of abbots and abbesses, long crumbled into dust, offer the most solid means of identification. These traces never amount to anything that might be deemed "normal" or "usual." In Cloisterham,

> so abounding [are the] vestiges of monastic graves, that the Cloisterham children grow small salad in the dust of abbots and abbesses, and make dirt-pies of nuns and friars; while every ploughman in its outlying fields renders to once puissant Lord Treasurers, Archbishops, Bishops, and such-like, the attention which the Ogre in the story-book desired to render to his unbidden visitor, and grinds their bones to make his bread.[16]

In the later Dickens works, replete as they are with pessimistic, weary visions of how the modern world is thwarted in its maturation by the past, such representations of history's traces are anything but reassuring.

Cloisterham is, in fact, an appropriate setting for Dickens's murder story because the idea of the past infiltrating the present suits the forensic work that the narrative and its major investigators need to do. The reasons for Edwin Drood's disappearance, whatever Dickens intended them to be, leave traces in Cloisterham Weir, two miles from the place where Edwin is last seen:

> No search had been made up here, for the tide had been running strongly down . . . , and the likeliest places for the discovery of a body, if a fatal accident had happened under such circumstances, all lay—both when the tide ebbed, and when it flowed again—between that spot and the sea. The water came over the Weir, with its usual

sound on a cold starlight night, and little could be seen of it; yet Mr. Crisparkle had a strange idea that something unusual hung about the place.

He reasoned with himself: What was it? Where was it? Put it to the proof. Which sense did it address? . . . Knowing very well that the mystery with which his mind was occupied, might of itself give the place this haunted air, he strained those hawk's eyes of his for the correction of his sight. He got closer to the Weir, and peered at its well-known posts and timbers. Nothing in the least unusual was remotely shadowed forth.[17]

The Weir runs through Canon Crisparkle's "broken sleep, all night," and he returns to the spot the following morning:

He had surveyed it closely for some minutes, and was about to withdraw his eyes, when they were attracted keenly to one spot. . . . It fascinated his sight. His hands began plucking off his coat. For it struck him that at that spot—a corner of the Weir—something glistened, which did not move and come over with the glistening water-drops, but remained stationary.

He assured himself of this, he threw off his clothes, he plunged into the icy water, and swam for the spot. Climbing the timbers, he took from them, caught among their interstices by its chain, a gold watch, bearing engraved upon its back E. D.

He brought the watch to the bank, swam to the Weir again, climbed it, and dived off. He knew every hole and corner of all the depths, and dived and dived and dived, until he could bear the cold no more. His notion was, that he would find the body; he only found a shirt-pin sticking in some mud and ooze.[18]

This episode appears to be equivocal on whether crime-scene evidence is *usual* or *unusual*. Like the Thames in "Night Walks," or the same river depicted in newspaper reports after the discovery of the unnamed man in the carpetbag, the river continues to flow, a symbol of the passing of time's usual current. Yet the presence of the gold watch, entangled with the Weir's timbers, is, like the carpetbag, unusual; it is a trace, a remnant of past activity, a ghost of an old play enacted on the same stage. It engenders Crisparkle's sense of the Weir as "haunted" and becomes ominously interwoven with his troubled sleep. The idea of a haunting takes us back to the abbots and abbesses and the Dickensian constructions of the past as peeping round the margins of the present. As with such phantoms, murder is a phenomenon that allows us to hear faint echoes of the past. Such an association is also painted in bold colors in *Bleak*

House when we read about a long terrace at Chesney Wold that is supposedly haunted by the ghost of a woman who was murdered, more or less, by her husband. "Let the Dedlocks listen for my step!" she utters moments before she dies, "and so sure as there is sickness or death in the family, it will be heard then."[19] Murder and its evidence are revenants similar to her ladyship's spectral footsteps throughout the work of Dickens: they become corroborations of the idea, writ large in forensic science as well, that extraordinary events leave echoes. In other words, medical jurisprudence's faith in the power of clues facilitated a belief in the significance of traces that is mirrored in Dickens's more imaginary vision of the enduring nature of human stories and the ghostly presence of the past.

Forensic science's interest in evidentiary traces was a self-reflexive investigation into the viability of objective truth—the "whole truth" conceit, in other words, on which the legal system based most of its solemn decisions. Medical jurisprudence's technologies of truth were often an analysis of interpretation and perception, a trend that is discernible in forensic science's increasing interest in photography both as a means of capturing patterns of behavior and as evidence of human peculiarities as unruly and multifaceted. In 1883 Francis Galton developed a composite form of photography that, he argued, "'brings into evidence' truths which are otherwise invisible to the eye."[20] Extending many of the arguments first broached in his earlier *Hereditary Genius* (1869), Galton reiterated his now famous view that, while variation is normal across populations and species, it is possible for the process of hereditary transmission to produce "types," or congenital subsets, who have a greater propensity for specific abnormalities and aberrant behaviors. As "moral and intellectual faculties" are "so closely bound up with the physical ones,"[21] he said, problems with mind and matter often affect the same people; like Cesare Lombroso, he believed that mental peculiarities are written into the physiognomies of individuals.[22] Yet, unlike his Italian counterpart, Galton did not provide specimens of "type" in numerous examples but combined a number of images into single composite portraits.

He hoped these composite portraits would produce a guide of "representative" types. With the help of his colleagues and correspondents, he collected photographic images of specific human subgroups, including the criminal and the sick. In order to identify a "central type" that indicated all that is good in a nation and revealed "the easiest direction in which a race can be improved,"[23] Galton superimposed "portraits like the successive leaves of a book," running images under a camera with a typically lengthy exposure time and thus producing a single composite image. "There can hardly be a more appropriate method of discovering the central physiognomical type of any race or group than that of composite portraiture," he claimed.[24]

In discussing Galton's use of photographs, Alexa Wright argues that "in their effort to establish a complete, objective and comprehensible social taxonomy and social order, Galton and his colleagues seem to have used photography as though it offered a direct equivalent to reality."[25] Indeed, ever since Daguerre and Fox Talbot unveiled their photographic apparatuses in 1839, the process of capturing likenesses on light-sensitive plates had promised a new dawn for realism in art and science. Edgar Allan Poe saw in early photographs a "perfect identity of aspect with the thing represented" and "truth itself in the supremeness of perfection."[26] In his important discussion of Galton's composite portraits in the context of the rise of forensic photography, Ronald R. Thomas perceives the "portrait of a 'criminal type'" to be an emphasis on the technology's "disciplinary powers."[27] Galton's "observing machine" made "the invisible visible" and, in its search for a representative "type," made "certain visible features disappear." What constitutes individual peculiarities, Thomas concludes, was "reduced to an insignificant blur."[28] This reading of forensic technologies in the nineteenth century seeks to explore the ways in which the new discourses surrounding the representation and control of criminality belonged to what Foucault had identified in *Discipline and Punish* (1975) as an era of new and complex mechanisms of power and control. The problem with this reading, for me, is that it reduces thoughtful moments of self-reflexivity in the works of men such as Francis Galton into an "insignificant blur," or a momentary deviation from the oppressive ideological projects driven by scientific thinkers of the past. Galton's perception of photography's "objectivity" appears to have been more complicated than both Thomas and Wright have suggested. Indeed, his writing reveals that he saw no blur as insignificant; his plan was not to make individual visible features disappear but to allow them to remain as traces—reminders of human variation and of the uneven mechanics of his photographic process. Galton writes, "It seems to me that it is possible on this principle to obtain a truer likeness of a man than in any other way." But he carefully avoids the absolute values implied in words such as "truth" and opts instead for "truer." Similarly, although his eugenic theories have often been associated with the "'racial purification' programmes of Nazi Germany,"[29] it seems that Galton would have rejected the idea of a pure or superior race because he believed, like his cousin Charles Darwin, in the living world as thriving through the "endless variety" caused by the "selective influences into close adaptations [organisms make with] their contemporaries, and to the physical circumstances of the localities they inhabit."[30] Although nature errs, as we see embodied in the criminal and pathological types, the benefits of variation mean that we should be concerned not with assimilating people into "a common type"[31] but in seeing a less circumscriptive and less definite "central type." Galton's composite portraiture "bring[s] into evidence all the traits in which there is agreement, and [leaves] but a ghost of a trace of individual peculiarities."[32] There is nothing pure about the images created; rather, their value resides in the fact that they do not remove or assimilate their ghosts but leave them as traces that testify to the peculiarities of their means of production. The same can be said of Galton's thoughtful writing. In the first paragraph of the volume, *Inquiries into Human Faculty and Its Development* (1883), in which composite portraiture is first outlined, he admits:

> I have revised, condensed, largely re-written, transposed old matter, and interpolated much that is new; but traces of the fragmentary origin of the work still remain, and I do not regret them. They serve to show that the book is intended to be suggestive, and renounces all claim to be encyclopaedic.[33]

This is a book that is interested in "the varied hereditary faculties of different men";[34] its objectives are found, then, among the peculiar, the ill-fitting, and the remarkable. Just as composite portraiture draws our attention to its modes of production and does not succeed in being or even seeking to be mimetic, Galton's treatise is shaped by the traces of its composite parts, and it "renounces" any drive toward objective authority.

The encyclopedic, like the objective, aims at a total, all-encompassing view, and Galton knows that blurs, gaps, and rude juxtapositions are "suggestive" of something less absolute. At this point, it is worth reminding ourselves of Pierre Macherey's *Theory of Literary Production* (1966), particularly the following passage:

> What begs to be explained in the work is not that false simplicity which derives from the apparent unity of its meaning, but the presence of a relation, or an opposition, between elements of the exposition or levels of the composition, those disparities which point to a conflict of meaning. This conflict is not the sign of an imperfection;

it reveals the inscription of an *otherness* in the work, through which it maintains a relationship with that which it is not, that which happens at its margins.[35]

This is certainly true of Galton's *Inquiries*' content as well as its form. Concerned with otherness in hereditary transmission and stitched together using a range of past sources, Galton's text contains images and narratives that lay bare what happens in the margins. The figure at the center of his pictures is typically surrounded by what Galton referred to as the ghost of an individual peculiarity. If we accept the claim that composite portraiture is a "truer likeness" of type than are other forms of representation, we gather a sense of how Galton's realism is very different from the potentially encyclopedic or objective modes of realism that he appears to have rejected. This is not a direct representation of the subject as it is in reality but a palimpsestic layering of representations that highlights not only shared characteristics but also the extraordinary traits that fall outside the borders of the usual.

In contrast to Thomas's focus on discipline and control, Nancy Armstrong has argued that the composite portrait shares with Dickens's literature a concern with veridical perception. Neither text nor photograph, she argues, "try to represent what was actually there . . . , even though both produce a remarkable density of visual detail to indicate that we are in the presence of something real." Galton, like Dickens, determines "the locations at which the multiplicity of urban phenomena will become visible for what they truly are."[36] And yet, in its interest in traces, blurs, and ghosts of individual peculiarities, composite photography, like Dickens's writing, betrays a more complex interest in the signifiers of what things *were*, or the traces left by the *processes* of interpretation rather than the thing itself: "To know the conditions of a work is to define the real process of its constitution, to show it is composed from a real diversity of elements which give it substance."[37] Dickens shares Galton's interest in the building blocks of representation and identification. For him, murder is an example of an individual peculiarity; it appears to haunt the composite portraits created by time's passing and shows the author to be interested in the characteristics and limitations of identification as an act as opposed to a statement of fact.

Richard D. Altick wrote some time ago that "the normally insignificant transactions of everyday existence" in the Victorian age were illuminated by murder as they would have been by the flash of a camera: "Witnesses, abruptly snatched from the usual obscurity of their lives, must recollect trivial circumstances which, had it not been for the fortuitous intrusion of a murder, would never again have figured in their memories."[38] Murders are useful to historians, Altick suggests, because they are extraordinary moments that shed light on the ordinary; they illuminate, lurid as it may well be, a way through an interminable labyrinth by the power of their extreme nature. Notwithstanding, the pictures described by Altick are very two-dimensional, or static: murder is the flash of a camera as Poe envisaged it—the correct capture of real life. Dickens, however, was regularly dismissive of naïve realism. In the first volume of *Household Words* he famously attacked John Everett Millais's 1849–50 painting *Christ in the House of His Parents* for its literalization of the New Testament:

> You will have the goodness to discharge from your minds all Post-Raphael ideas, all religious aspirations, all elevating thoughts; all tender, awful, sorrowful, ennobling, sacred, graceful, or beautiful associations; and to prepare yourselves as befits such a subject—Pre-Raphaelly considered—for the lowest depths of what is mean, odious, repulsive, and revolting.[39]

Although Dickens criticized the Pre-Raphaelite Brotherhood for its wish to copy its subjects with "utmost fidelity,"[40] he appears to have been more offended by Millais's representations of ugliness: Christ is "a hideous, wrynecked, blubbering, red-headed boy"; the Virgin Mary is "so horrible in her ugliness, that . . . she would stand out from the rest of the company as a Monster, in the vilest cabaret in France, or the lowest gin-shop in England"; and "the carpenters might be undressed in any hospital where dirty drunkards, in a high state of varicose veils, are received."[41] His opposition was not pitched against Millais's wish to be truthful, then, but rather his way of forgetting, with his keen attention to naturalistic detail, art's duty to enrich, beautify, and educate. "The regulation of social matters, as separated from the Fine Arts," he satirically concludes, "has been undertaken by the Pre-Henry the Seventh Brotherhood."[42] Millais had missed the opportunity to explore how interpretation and identification work in forms that often are more dynamic than might be attained by veridical representation.

In a letter to John Forster prior to the establishment of *Household Words*, Dickens famously reported that he saw his new journal as behaving like "a certain SHADOW, which may go into any place, by sunlight, moonlight, starlight, firelight, candlelight, and be in all homes, and all nooks and corners, and be supposed to be cognisant of everything, and go everywhere, without the least difficulty . . . , a kind of semi-omniscience, omnipresent, intangible creature."[43] "Certain SHADOW" emphasizes the noun in a way that draws our attention from the more interesting and abstruse use of "certain," which is a word that can mean an object identified as known yet unnamed, on the one hand, or absolutely sure, on the other hand. There is a lack of commitment to reliable meaning in the word "SHADOW" too, as the signified is defined, conflictingly, as both omnipresent and *semi*-omniscient—an invitation, we might argue, to doubt the prospect of complete and absolute knowledge. Dickens preferred instead to go astray, to have his stories take ghost walks and to make his characters dream, intuit, and half remember. In Mr. Crisparkle's search of the Cloisterham Weir, knowledge is embodied in the discovery of clues to a murder mystery. The scene in which this occurs violates any sense of realism by having Crisparkle feel the Weir to be haunted; he also has strange intuitions that evidence will be found at the scene, and his discovery of the jewelry is simply an impossible coincidence. The conceit of murder allows the novel to embody as well as to demonstrate how interpretation involves looking for traces of what exists in and is counter to what is directly given to us in terms of perception.

Take the scene in *Martin Chuzzlewit* (1843–44) in which the roguish Montague Tigg is murdered by Jonas Chuzzlewit. The episode sheds light on three major subjects: ordinary life incidental to the murder plot (thus supporting Altick's point), how narrative (such as in a murder story) is constructed, and, most complexly, how murder interpretation works self-reflexively in the margins and often in conflict with the representative structures it is made up of. To begin with the first and most simple of these points, when Chuzzlewit travels to the place where he will kill Montague, the narrative prefigures a point Dickens would later make about the chopped-up murdered man sleeping soundly in his bed, unaware of what is to come:

> Did no men passing through the dim streets shrink without knowing why, when he came stealing up behind them? As he glided on, had no child in its sleep an indistinct perception of a guilty shadow falling on its bed, that troubled its innocent rest? Did no dog howl, and strive to break its rattling chain, that it might tear him; no burrowing rat, scenting the work he had in hand, essay to gnaw a passage after him, that it might hold a greedy revel at the feast of his providing? . . . The fishes slumbered in the cold, bright, glistening streams and rivers, perhaps; and the birds roosted on

the branches of the trees; and in their stalls and pastures beasts were quiet; and human creatures slept.⁴⁴

The scene is conspicuous for the seeming lack of a response to the imminent, illuminating melodrama passing through it. Falling asleep himself, Jonas dreams of a "great crowd" filling a street and looks forward to what Galton would create in his composite portraits:

> [He] stood aside in a porch, fearfully surveying the multitude; in which there were many faces that he knew, and many that he did not know, but dreamed he did; when all at once a struggling head rose up among the rest—livid and deadly, but the same as he had known it—and denounced him as having appointed that direful day to happen.⁴⁵

Just as the ghost of an individual's peculiarity in Galton's forensic photographs parallels how murder allows moments to stand out in history, so this livid and deathly imagination of the victim's head draws attention to the one "direful day" as an appointed, marked, and recognized point in time. Finally, once Jonas has killed Montague in a wood, he is plagued, as Bill Sikes is, by his fearful and guilty conscience:

> He had had a terror and dread of the wood when he was in it. . . . Dread and fear were upon him, to an extent he had never counted on, and could not manage in the least degree. He was so horribly afraid of that infernal room at home. This made him, in a gloomy murderous, mad way, not only fearful *for* himself, but *of* himself; . . . he became in a manner his own ghost and phantom, and was at once the haunting spirit and the haunted man.⁴⁶

Irrationally, he thinks, on reaching his home:

> What if the murdered man were there before him!
>   He cast a fearful glance all round. But there was nothing there. . . . Looking in the glass, [he] imagined that his deed was broadly written in his face, and lying down and burying himself once more beneath the blankets, heard his own heart beating Murder, Murder, Murder, in the bed; what words can paint tremendous truths like these!⁴⁷

"The Ghosts of Individual Peculiarities"

Tremendous truths fall outside the borders of representation, just as Jonas's guilt lands outside any realist impulse in the novel. His guilt becomes a haunting in the sense that it reminds us how murder is a ghostly trace, unobscured by the layering processes of time and history. Jonas's conscience is, in fact, a self-conscious fictionalization—a moment in which the narrative refers to its own artifice. Whereas most of the guilty imaginings are filtered through the murderer's troubled fancy, the conceit of his heart beating "Murder, Murder, Murder" is unqualified by anything that would allow it to sit comfortably in a realist account. Shortly before this passage was written, Poe had published "The Tell-Tale Heart" (1843), a story Dickens was likely to have read, given his admiration for its author and the fact that he had met Poe during his first trip to America in 1842.[48] In a way that seems obvious, Dickens's representation of Jonas's guilty heart could belong to one of Poe's gothic fantasies; it certainly shares Poe's gothic heritage. Thus, while the murder of Tigg provides a glimpse into the ordinariness of a specific moment before the killing and simultaneously highlights how homicide is a spectral trace that permits a historical moment to be distinguished, its Gothicism works on the margins of that constructive process itself. Developed through forensic reconstruction, Dickens's narrative style violates the realist method enough to provide the distance it needs to question what works and what fails as interpretation, identification, and reconstruction.

To conclude with the example with which I began, in "Night Walks," Dickens describes a walk past Bethlehem Hospital that he took after he had crossed Waterloo Bridge and remembered the man in the carpetbag. By this point he has also visited an abandoned theater and Newgate Prison; the lunatic asylum is the natural next stop, therefore, in his morbid night pilgrimage:

> I had a night-fancy in my head which could be best pursued within sight of its walls and dome. And the fancy was this: Are not the sane and the insane equal at night as the sane lie a dreaming? Are not all of us outside this hospital, who dream, more or less in the condition of those inside it, every night of our lives? . . . Do we not nightly jumble events and personages, and times and places, as these do daily? . . . Said an afflicted man to me, when I was last in a hospital like this, "Sir, I can frequently fly." I was half ashamed to reflect that so could I by night. . . . I wonder that the great master who knew everything, when he called Sleep the death of each day's life, did not call Dreams the insanity of each day's sanity. (153–54)

The central message here is that it is impossible to know where truth and irrationality begin, end, and overlap. Just as each day's sanity has its insanity, so the ordinariness of each day has its extraordinariness, and nowhere is this more obvious, I argue, than in murder. Homicide is the exaggeration, the drama, and the scandal of historical perspective; it sheds light both on the quotidian and on the past. But this enlightenment is not simply a matter of opening a door and allowing one and all to see the truth; it is a means of questioning, in the act of imagining, *how* we see. Indeed, as the story of a perception that was both penetrative and hopeless, the story of the Waterloo Bridge murder suited Dickens's imaginary realism as a symbol of how interpretation is always limited by its creative nature. Dickens's "concern with crime was . . . more persistent and more serious than most men's"[49]—not only, as Philip Collins implies, by virtue of crime's importance as a sociological issue but because it was a means of exploring the strategies we have—specialist, fictional, and everyday—for performing acts of identification.

Notes

1. "Supposed Murder: Another Account," *The Times*, 10 October 1857, 9.
2. Ibid.
3. Ibid.
4. Ibid.
5. "The Waterloo-Bridge Murder," *The Times*, 27 October 1857, 11.
6. Ibid.
7. Ibid.
8. "Supposed Murder," 9.
9. Charles Dickens, "Night Walks," in *Dickens's Journalism*, vol. 4, "*The Uncommercial Traveller*" *and Other Papers, 1859–70*, ed. Michael Slater and John Drew (story originally published 1860; London: Dent, 2000), 151.
10. Ibid., 150.
11. Ibid., 151.
12. John Bowen, *Other Dickens: Pickwick to Chuzzlewit* (Oxford: Oxford University Press, 2000), 5.
13. Quoted in Philip Collins, *Dickens and Crime* (1962; London: Macmillan, 1965), 1.
14. Charles Dickens, *The Mystery of Edwin Drood*, ed. Margaret Cardwell (1870; Oxford: Oxford University Press, 1982), 1, 12.
15. Ibid., 13.
16. Ibid., 12–13.
17. Ibid., 144–45.
18. Ibid., 145.
19. Charles Dickens, *Bleak House*, ed. Andrew Sanders (1852–53; London: Dent, 1994), 85.
20. Ronald R. Thomas, *Detective Fiction and the Rise of Forensic Science* (Cambridge: Cambridge University Press, 1999), 125.
21. Francis Galton, *Inquiries into Human Faculty and Its Development* (London: Macmillan, 1883), 4.
22. See Thomas, *Detective Fiction*, 123.
23. Galton, *Inquiries*, 14.
24. Ibid., 9, 15.
25. Alexa Wright, *Monstrosity: The Human Monster in Visual Culture* (London: I. B. Tauris, 2013), 68.

26. Edgar Allan Poe, "The Daguerreotype," *Alexander's Weekly Magazine*, 1840, quoted in Thomas, *Detective Fiction*, 111–12.
27. Thomas, *Detective Fiction*, 123.
28. Ibid., 123, 126.
29. Ibid., 67.
30. Galton, *Inquiries*, 3.
31. Ibid.
32. Ibid., 10.
33. Ibid., 1.
34. Ibid.
35. Pierre Macherey, *A Theory of Literary Production*, trans. Geoffrey Wall (1966; London: Routledge, 2006), 89. Italics in original.
36. Nancy Armstrong, *Fiction in the Age of Photography: The Legacy of British Realism* (Cambridge, MA: Harvard University Press, 1999), 147–48.
37. Macherey, *Literary Production*, 56.
38. Richard D. Altick, *Victorian Studies in Scarlet* (New York: W. W. Norton, 1970), 13.
39. Charles Dickens, "Old Lamps for New Ones," *Household Words* 1 (1850): 265.
40. Ibid., 266.
41. Ibid., 266–67.
42. Ibid., 267.
43. John Forster, ed., *The Life of Charles Dickens* (London: Chapman and Hall, 1876), 2:78–79.
44. Charles Dickens, *Martin Chuzzlewit*, ed. Michael Slater (1843–44; London: Dent, 1994), 678.
45. Ibid., 679.
46. Ibid., 681–82, 684.
47. Ibid., 685.
48. Michael Slater, *Charles Dickens* (London: Yale University Press, 2009), 184.
49. Collins, *Dickens and Crime*, 1.

References

Altick, Richard D. *Victorian Studies in Scarlet*. New York: W. W. Norton, 1970.
Armstrong, Nancy. *Fiction in the Age of Photography: The Legacy of British Realism*. Cambridge, MA: Harvard University Press, 1999.
Bowen, John. *Other Dickens: Pickwick to Chuzzlewit*. Oxford: Oxford University Press, 2000.
Collins, Philip. *Dickens and Crime*. London: Macmillan, 1965. Originally published 1962.
Dickens, Charles. *Bleak House*. Edited by Andrew Sanders. London: Dent, 1994. Originally published 1852–53.
———. *Martin Chuzzlewit*. Edited by Michael Slater. London: Dent, 1994. Originally published 1843–44.
———. *The Mystery of Edwin Drood*. Edited by Margaret Cardwell. Oxford: Oxford University Press, 1982. Originally published 1870.
———. "Night Walks." In *Dickens's Journalism*, vol. 4, *"The Uncommercial Traveller" and Other Papers, 1859–70*. Edited by Michael Slater and John Drew, 148–57. London: Dent, 2000. Story originally published 1860.
———. "Old Lamps for New Ones." *Household Words* 1 (1850): 265–67.
———. *Oliver Twist*. Edited by Stephen Gill. Oxford: Oxford University Press, 2008. Originally published 1837–39.
Forster, John, *The Life of Charles Dickens*. 2 vols. London: Chapman and Hall, 1872–74.
Galton, Francis. *Inquiries into Human Faculty and Its Development*. London: Macmillan, 1883.
Lawson, Kate, and Lynn Shakinovsky. *The Marked Body: Domestic Violence in Mid-Nineteenth-Century Literature*. Albany: State University of New York Press, 2002.
Macherey, Pierre. *A Theory of Literary Production*. Translated by Geoffrey Wall. London: Routledge, 2006. Originally published 1966.
Matchett, Willoughby. "The Chopped-Up Murdered Man." *The Dickensian* 14 (1918): 117–19.

Slater, Michael. *Charles Dickens*. London: Yale University Press, 2009.

Surridge, Lisa. *Bleak Houses: Marital Violence in Victorian Fiction*. Athens: Ohio University Press, 2005.

Thomas, Ronald R. *Detective Fiction and the Rise of Forensic Science*. Cambridge: Cambridge University Press, 1999.

The Times. "Supposed Murder: Another Account." 10 October 1857, 9.

The Times. "The Waterloo-Bridge Murder." 27 October 1857, 11.

Trodd, Anthea. *Domestic Crime in the Victorian Novel*. London: Macmillan, 1989.

Tromp, Marlene. *The Private Rod: Marital Violence, Sensation, and the Law in Victorian Britain*. Charlottesville: University Press of Virginia, 2000.

Wright, Alexa. *Monstrosity: The Human Monster in Visual Culture*. London: I. B. Tauris, 2013.

4.
"A Puzzle of Character"
Francis Iles and Narratives of Criminality in the 1930s

Victoria Stewart

The adoption of a pseudonym is often a way for authors to mark a divergence from a kind of writing with which they have become associated. Recent examples include John Banville, who uses the pseudonym Benjamin Black when writing detective fiction, and J. K. Rowling, who has published crime novels under the name Robert Galbraith. In the interwar period, authors also used alternative names as a means of managing reader expectations and, implicitly, to signal generic distinctions. In some cases, the fact that the works published under the different names were written by the same individual was deliberately kept concealed from the reading public until "revealed" by the press: Laura Thompson notes that although Agatha Christie began publishing psychologically focused, often autobiographical novels as Mary Westmacott in 1930, it was only in 1949 that the *Sunday Times* confirmed that Christie and Westmacott were one and the same.[1] These doublings, then, underline the importance of recognizable names in the process of genre formation. But in the case of the subject of this chapter, who was born Anthony Berkeley Cox, questions of identity run deeper and are inherent in the generic experiment that the use of a pseudonym facilitated.

Cox used his birth name at the very start of his writing career in the 1920s, but it was as Anthony Berkeley that he initially established himself as a writer of detective novels, nonfictional crime writing, and book reviews. In 1931, he published *Malice Aforethought* under the name Francis Iles, which he took from an ancestor on his mother's side who was, apparently, a notorious smuggler.[2] This novel and two subsequent works by Iles, *Before the Fact* (1932) and *As for the Woman* (1939), were widely considered at the time to be breaking new ground in the detective fiction genre, but Iles, like Berkeley, was not restricted to writing fiction and also produced nonfictional crime writing during the 1930s. From the late 1930s until his death in 1971, the author's output consisted largely of book reviews. As Francis Iles, he was a regular reviewer of detective fiction for the *Sunday Times* from 1952 until 1956 and for the *Manchester Guardian* from 1957 until 1970. As Malcolm Turnbull notes, Berkeley enjoyed the "speculation about Iles's true identity" that followed the publication of *Malice Aforethought*, and it was only in the 1950s that he acknowledged publicly that the works of Berkeley and of Iles were by the same person.[3] In what could be read as either a mischievous or a deliberately obfuscatory move, *Great Unsolved Crimes*, a 1938 essay anthology, has chapters by both Iles and Berkeley; similarly, the radio series *Connoisseurs of Crime*, broadcast on the BBC's Home Service in the late 1950s, includes one episode in which Iles and Berkeley collaborate and Anthony B. Cox is listed as having composed the music.[4]

In considering the nature and reception of the generic experiment undertaken in the Iles novels, which will be my focus here, issues of the slipperiness of identity are crucial and operate not only in relation to the testing of genre boundaries but also on a thematic level. Iles's novels are peopled with individuals, particularly men, who present themselves as one thing but are actually another. In this regard, the novels engage with questions of identification that were pressing during the 1930s. The shifting social structures of the post–Great War period presented new opportunities for social mobility but prompted anxiety that individuals might lay claim to social positions to which they were not entitled. Meanwhile, although popularized versions of the discourse of psychoanalysis seemed to offer a key to understanding identity, they also revealed potentially uncontrollable impulses beneath the surface of civilized behavior, an issue that was particularly pertinent to contemporary efforts to define and control criminality. Traditional detective fiction characterized the identification of the criminal as a logical rather than psychological problem, and in focusing on the criminal, Iles questions how genres are constituted and brings wider concerns about the identification of criminality into play.

Contextualizing Iles's first novel, *Malice Aforethought*, in relation to debates about detective fiction in the 1930s can help explain why his works were considered noteworthy, quite aside from the mystery attached to their authorship. In the typical detective novel of the post–Great War period, a crime occurs at or before the start of the narrative, with the focus thereafter resting on the identification of the culprit, a process that itself involves reconstructing events and relationships from the period before the crime. Such novels evidence the "duality" that Tsvetan Todorov identifies between the "story of the crime and the story of the investigation.... The first story, that of the crime, ends before the second begins."[5] This model explains how crime in such texts can be presented as an intellectual puzzle, with the importance of psychological motivations diminished, as motives are constructed more often in material than in emotional terms. Shifting attention to the culprit, as Iles does in *Malice*, necessitates a fuller engagement with the question of motive and offers a potentially much more troubling object of identification for the reader, who is asked to follow the thought processes not of the detective, set on righting wrongs, but of the criminal, intent on committing crime. This change in emphasis also means that the retrospective piecing together of past events tends to be displaced by a focus on the sequential, future-oriented narrative trajectory that is one of the key generic markers of the thriller, a genre that typically centers on a solitary hero battling not against a single enemy but against a conspiracy. Whereas the hero of the thriller is usually constructed as a force for good (even if we might question the essentially conservative ideology underpinning many of these narratives), the protagonists of the novels to be discussed here are at best highly ambivalent.

It might be presumed that introducing psychology into the crime novel would equate to greater realism and that the result would be a narrative resembling, for instance, Fyodor Dostoyevsky's *Crime and Punishment* (1868). But meditations on crime of a protoexistential kind are less important here than the effect that shifts in perspective—or more specifically, focalization—have on structure. Moving away from the depiction of crime as a kind of puzzle, Iles's novels produce instead the tension concomitant with following a protagonist from one perilous situation to the next. These works are on the threshold between the understanding of crime as a social problem and its pathologization: the affectless, deranged, or motiveless killer is a figure who cannot be contained within the parameters of classic detective fiction, and Iles's works foreshadow the emphasis on the psychology of the criminal to be found in the writings of post–World War II authors such as Patricia Highsmith. But Iles is at pains to root these criminals in specific sociocultural contexts, producing a heightened sense of their engagement with post–Great War understandings of gender relations and, especially, masculinity. Further, while each of the Iles novels brings the culprit into focus in a way that is untypical of detective fiction during this period, *Before the Fact* goes further and is focalized by the eventual victim, a choice that not only challenges further the ways in which readers might position themselves in relation to the protagonists but also provides Iles with the opportunity to construct a highly self-reflexive narrative.

Challenging Generic Conventions

Classic detective fiction is not usually presumed to have much to do with either real-life crime or literary realism, although as Gill Plain points out, there is actually something disingenuous about the notion of the detective novel as "pure puzzle": "Any intellectual 'purity' is always already compromised by the figure of the detective, who must be embodied, and with embodiment comes implication within the social."[6] Commentators on the form in its heyday grappled in various ways with this issue. Iles's friend and fellow detective novelist Dorothy L. Sayers was criticized for giving too much prominence to the romance between her investigating protagonists, Lord Peter Wimsey and Harriet Vane, in her novels of the 1930s. Sayers was herself a prolific commentator on detective fiction and by the early 1930s had come to the view that an increasing engagement with realism would ultimately enrich the form. Writing in 1931, she mentioned *Malice Aforethought* as an example of "the new interest in criminal psychology," identifying such works more as "studies in murder than detective-stories";[7] C. S. Forester, as Sayers notes, was undertaking similar experiments during this period. Sayers's desire to coin a new name for this apparently new form is itself telling—genre labels are capacious up to a point, but beyond this point a new designation is required. Even the supposedly more conventional works of Anthony Berkeley show an awareness of these shifts. Sayers quotes the preface of *The Second Shot* (1930), a novel featuring his series detective Roger Sheringham, in which Berkeley predicts that, as far as the future of the detective form is concerned, "the puzzle will remain but it will become a puzzle of character."[8] Thus, although Sayers is careful to identify Iles's novels as something other than detective fiction, there is a tentative acceptance, on both her part and Berkeley's, that all detective fiction could and perhaps should incorporate new, extraliterary understandings of individuality and, specifically, criminality.

Elsewhere, however, Sayers identifies the potential problems with this shift, and her concerns were shared by other commentators who were wary of the form's potential evolution. In "Aristotle on Detective Fiction" (1935) she lamented the prevalence of novels with "a rather slender plot" and a "good deal of morbid psychology."[9] For Sayers, diverging from normative psychology is problematic because it requires overly specialized knowledge on the part of the reader, and it produces "books of a certain psychological interest . . . but remote from the thought-processes of normal humanity."[10] On this point, Sayers is in agreement with H. Douglas Thomson, who, in one of the earliest monographs on the form, deplored the "tendency to make the detective story a psychological study." Preferring detective fiction that laid no claim to realism, he deplored the recent tendency of the form to "borrow its characters from the sciences."[11] These objections reflect Thomson's adherence to the idea that the reader should have as much a chance of finding the solution to the puzzle as the detective: appealing to psychological analyses of character is, by this account, equivalent to expecting the reader to recognize the effects of obscure poisons. But Sayers's references to "morbid psychology" emphasize that psychology tends to come into focus in fiction only when it relates to abnormality. In classic detective fiction, motive is almost inevitably understood in material and social terms; money and preservation of social status are frequent motives, which the reader is presumed to understand if not endorse, and there is no space for the psychologically confused killer. Objections to psychology in detective fiction emerge not just from a sense that unpredictable characters spoil the reader's fun: acknowledging how crime is understood in disciplines such as psychology reveals that attempts to sequester detective fiction from its social context through the consideration of its apparently ahistorical aspects, such as the "puzzle" structure, are doomed to fail. These debates about the boundaries of genre are underpinned by a classificatory urge, a desire to identify and anatomize particular kinds of literature and their subsets, that is at odds with the often unruly nature of both the texts and the actions they purport to depict.

In addition to this generalized understanding of psychology as a way of explaining motives by reference to thoughts and feelings rather than actions, a related and equally generalized understanding of psychoanalysis was also emerging in literature and literary culture at this time. The ideas that individuals have repressed thoughts and wishes that may express themselves in unexpected ways and that childhood and attitudes toward sex and sexuality are both crucial in individual development are manifested in literature of the 1920s and 1930s in what Nicola Humble refers to as "Freudianism without Freud."[12] She suggests that part of the reason why psychoanalytic and psychological discourse was particularly problematic for detective fiction was that it "misse[d] . . . the profound individuality of each crime and criminal."[13] This might seem ironic given the extent to which detective fiction deals with types, but detective novels are concerned with deducing motive and opportunity from immediate material circumstances and with the formation of a plot of "ingenious originality" rather than with debating the causes of habitual criminality or the individuating psychological factors that might lead to a crime being committed.[14] Focalizing the narrative from the perspective of the criminal, however, necessitates a confrontation with precisely these issues.

"No One Quite Knew What Johnnie Was": Shifting Social Structures in Iles's Novels

Iles's novels engage with the middle- and upper-middle-class world familiar from much classic detective fiction. Typically, the action of these novels unfolds in what W. H. Auden termed a "closed world," which might be a school, a workplace, or, as in Iles's case, a village. In being closed, these settings place restrictions on the number of possible suspects.[15] Individuals may have secrets, quite aside from criminality, that they wish to keep concealed, and their attempts to retain a facade of respectability can cause complications in the unraveling of the crime plot. In both *Malice Aforethought* and *Before the Fact*, Iles exposes from the outset the "dark underbelly" of apparently respectable social worlds and, in the process, both provides a glimpse of the successes and failures of interwar social mobility and offers supposedly scientific explanations of his protagonists' behavior.

The opening sentence of *Malice* announces Bickleigh as a potential murderer: "It was not until several weeks after he had decided to murder his wife that Dr Bickleigh took any active steps in the matter."[16] The subsequent chapter reveals that he is bullied by his wife but also that he avails himself of the relative freedom of movement offered by his role as a doctor to indulge in prolific womanizing. What follows in the second chapter is essentially a "case-study" of Bickleigh, one that combines social and psychological explanations for his self-perception. A chemist's son who attends "one of the very minor universities," he becomes a doctor and, on taking up the practice in Wyvern's Cross that his father has "bought for him with the last of his savings and a mortgage," discovers that "birth [counts] for everything and achievement nothing."[17] But over and above these issues of social status, with self-improvement evidently condemned as social climbing, the narrator emphasizes the significance for Bickleigh of his slight build:

> The smallness of Dr Bickleigh's stature was responsible for almost everything that he then was. . . . Physical appearance plays a larger part in the formation of character than is always recognized. . . . In these days of glib references to complexes, repressions, and fixations on every layman's lips, it is not to be supposed that Dr Bickleigh did not know what was the matter with him. He could diagnose an inferiority complex, and a pronounced one at that, as well as anyone else.[18]

The tone here is slippery: the narrator adopts a quasi-essayistic perspective, both appealing to and keeping at arm's length the discourse of psychoanalysis. The narrator seems to offer social and psychological explanations as a sop, however: the real interest is in the twists and turns of the plot. Iles's use of irony passes moral adjudication back to the reader, who is left with the realization that the social, psychological, and legal constructions of Bickleigh that emerge over the course of the narrative are of little help in reaching such a judgment.

The lineage of Johnnie Aysgarth in Iles's *Before the Fact* is more "respectable," though the explanations for his behavior are hazy, not least because the narrative is focalized from the perspective of his wife, Lina, rather than from Johnnie himself. Johnnie is one of the four sons of Sir Thomas Aysgarth: "one had been killed in the war, one was in Australia, nominally sheep-farming, one was on the stage, and Johnnie, the youngest, was—well, no one quite knew what Johnnie was."[19] This throwaway sketch of the interwar decline of the minor nobility can be contrasted with the hard climb to social respectability undertaken by Bickleigh: Bickleigh's acquired social status will never make up for his lack of lineage, while Johnnie can continue trading on his birth despite being penniless, a predicament exacerbated by the fact that he is an inveterate gambler as well as a womanizer and, as Lina eventually comes to realize, a murderer. Johnnie's psychology is not anatomized in the same systematic case-study fashion as is Bickleigh's. Instead, the interactions that Lina observes and of which she is a part reveal him to be a barefaced liar who manages to humiliate his wife into compliance. *Before the Fact*—like both *Malice Aforethought* and Iles's third novel, *As for the Woman*—now reads as a narrative of domestic abuse; Lina is belittled and manipulated by her husband but finds herself inexorably drawn back into his clutches, apparently becoming, in the final pages of the novel, a willing murder victim.

Whereas *Malice* makes no secret, from the very opening line, that Bickleigh is a potential murderer, the tension in *Before the Fact* comes from doubt. For much of the novel, Lina is able to persuade herself that Johnnie's actions have "innocent" explanations. He also on occasion brazenly admits wrongdoing but in such a way as to make her appear prudish or naive for objecting. When she is upset while they are on honeymoon, upon realizing Johnnie has conned a waiter out of fifty francs, he laughs the incident off: "You funny little thing! Aren't you delighted at having put one over on those thieves for a change?"[20] There is doubt, then, over the course of the narrative as to whether Lina is right to suspect Johnnie or just paranoid; for the reader, though, it is a case of attempting to deduce not the fact but the scope of his criminality. The idea that Lina might be overinterpreting Johnnie's behavior is played out explicitly in Alfred Hitchcock's adaptation of the novel, *Suspicion* (1941), which for the most part follows the novel's trajectory but changes the ending. The novel ends with Lina lying in bed, waiting for her husband to bring her what she has convinced herself will be a poisoned drink: "It did seem a pity that she had to die, when she would have liked so much to live."[21] Hitchcock ended the film with Johnnie no longer under suspicion and the couple united; indeed, Lina apologizes to him for having suspected him in the first place. This alternative ending seems to acknowledge the narrative logic of the traditional detective novel in which clues are planted that lead the reader in different directions and potential alternative solutions are held in balance until the point at which the detective finally identifies the culprit. In theory, a different outcome could have been deduced from the same clues. Rather than a range of suspects, however, we are presented in both the novel and the film with a single individual whose actions can be interpreted in two opposing ways. This is ostensibly the case, at least: part of the reason why the ending of *Suspicion* can seem unsatisfactory is that even if Johnnie is not a murderer he has still victimized and exploited Lina.

It is striking, however, that even when focalizing a narrative from the perspective of a potential, indeed eventual, victim, Iles still uses third-person narration to keep her at arm's length from the reader rather than promote the reader's identification with her. Leroy Lad Panek goes so far as to state that "any reasonable reader who experiences *Before the Fact* wants to strangle [Lina] for her abysmal stupidity even before she is murdered at the end of the book."[22] This comment illustrates the effects that Iles's technique can have at its most extreme: in his attempt to challenge conventions of both plotting and character and despite appealing to recognizable psychological concepts and ideas, he ends up producing novels that apparently allow for less engagement with their protagonists than does even the much-maligned clue-puzzle form. But if Iles takes a sometimes self-satisfied delight in pulling the rug out from under the feet of the reader, this shiftiness can itself be seen as symptomatic of the social world of the 1930s in which the narratives are rooted. The codes of "respectability" that ought to help in identifying individuals and placing them within the social order are revealed to be unstable and all too easily mimicked.

Making a Murderer

Aside from the tangled network of social and psychological influences that underpin his protagonists' actions, Iles is keen to emphasize another factor that affects their self-perception and behavior: their reading. If references to psychological concepts complicate the idea of the detective-novel world as a self-contained one, references to what the protagonists read also serve to break the frame of the narrative. Thematizing reading can work either to underscore the fictiveness of the narrative or to inject a further degree of realism; in either case it also suggests that reading may be an act that has consequences in the world. The idea of detective fiction as an intellectual exercise is at odds with the implication that reading about crime, which often features in these texts, might actually provide models of criminal behavior.

In Bickleigh's case, the first step to murder occurs when he uses a sample of patent medicine to give his wife Julia a headache so that she will take to her bed and not bother him: "He had seen [the medicine] reviewed . . . in the *BMJ*. . . . It had been a failure . . . because . . . it induced such violent headaches."[23] He repeats this tactic, and eventually, when she complains of the severity of her symptoms, he injects her with morphine as a remedy. She becomes addicted to this opiate and administers the fatal dose herself when Bickleigh, having warned her about the potential consequences, leaves a syringe out ready for her, knowing she will be desperate enough to take it. Having apparently succeeded in carrying out an undetectable murder—effectively an engineered suicide—Bickleigh's confidence is further nurtured by his reading: "Dr Bickleigh read de Quincey on *Murder as a Fine Art*. . . . On the whole . . . he agreed with the author. Murder could be a fine art: but it was not for everyone. Murder was a fine art for the superman. It was a pity that Nietzsche could not have developed de Quincey's propositions."[24] Bickleigh's belief that he may indeed be a "superman" is framed by Iles as a source of comedy; such a self-perception is a ridiculous one for a doctor in a country backwater to hold. Nevertheless, after orchestrating the death of his wife in order to pursue a relationship with their neighbor Madeleine, Bickleigh goes on to unsuccessfully attempt to poison his supposed rival for Madeleine's affections. In having Bickleigh play out his overliteral understanding of his reading matter, Iles references both the idea of the "undetectable" murder and the concept of the murderer who sees himself as superior to ordinary individuals and therefore entitled to act on his desires. These ideas would have been familiar to contemporary readers. The 1924 trial, widely reported in Britain, of Nathan Leopold and Richard Loeb, the Chicago students found guilty of the murder of eleven-year-old Bobby Franks, condensed these two themes and saw Nietzsche cited as an influence on the defendants' behavior.[25] Being able to devise in literary form a crime that cannot be detected is one thing, but the idea that there may be murderers out in the real world going unpunished because they have used precisely such techniques to evade detection is quite another, and keeping them separate is a delicate procedure.

Before the Fact also engages explicitly with the concept of the undetectable murder. As Patrick Faubert notes, for much of the novel, Lina is "cast as an unwitting detective."[26] Having realized what her husband has planned for her, Lina becomes complicit in his crime to the extent of writing a note to be opened after her death, indicating that she has committed suicide. But complicity is perhaps not the correct way of describing the process by which Lina succumbs to being a victim. During the final fifth of the novel, she reacts to Johnnie in the way that a reader might when reaching a particularly suspenseful part of a novel. She wants him to act simply to relieve her own sense of dread, even if his actions will result in her own death: "Things could not go on like this. Lina began seriously to wonder whether she would not actually let Johnnie kill her, and put an end to it all."[27] We are told early on in the novel that "Johnnie reads nothing but detective stories," and this quasi-metafictional conceit is echoed in Johnnie's interactions with Isobel Sedbusk, a crime novelist whom he persuades to tell him about the "undetectable poison" that he will use on Lina.[28] Sedbusk lends Johnnie a factual book about notorious Victorian poisoner William Palmer.²⁹ Lina, meanwhile, borrows from her another volume, "a penetrating piece of work about murders and murderers. Analysing her subject, the authoress had suggested that just as there are born murderers so there are born victims, murderees, whose natural destiny it is to get murdered."³⁰ Iles is evidently referencing F. Tennyson Jesse's book *Murder and Its Motives* (1924), which alludes to the idea of the fated victim, or "murderee," a concept that, as I have discussed elsewhere, is clearly even more problematic than the notion that there might be born murderers.³¹ Once Lina accepts the role of fated victim, there is no hope for her.

Notably, in a dinner table conversation that reveals the persistence of the belief that appearance might dictate destiny, Sedbusk declares that Johnnie "couldn't commit murder if [he] tried for a hundred years."³² This comment was echoed in the reactions of audience members at test screenings of *Suspicion*, who were unable to believe that a character played by Cary Grant, an actor associated with screwball comedies, "would turn out to be truly evil."³³ While Johnnie's eventual actions and, even at this point in the narrative, the reader's suspicions about what he may have already done cast a shadow of irony over Sedbusk's comments, there is a connection here to the debates that emerge in Iles's work about how the boundaries between legality and morality might be drawn. Where the deaths of their respective wives are concerned, both Johnnie and Bickleigh require the acquiescence or the active participation of their victims, and psychological pressure rather than physical action is the key. Sedbusk's comment allows Johnnie to believe that he is not and could never be a murderer because of how he has orchestrated his crime and because, in his own view, his actions transcend legal and moral niceties.

The murder—or rather, murder attempt—in *As for the Woman* is somewhat different and could be placed at the impulsive end of the scale, but it too can be considered as an adjunct to a complex and troubling depiction of masculinity. In this novel, the focalizing protagonist is a young man, Alan, who, after a patch is discovered on his lung, is sent away to convalesce at the home of Dr. Fred Pawle and his wife, Evelyn. Alan is frequently infantilized by the couple, compounding the emasculatory implications of his illness. This device echoes *Before the Fact*, in which Johnnie is often described as being immature or acting like a boy. Unlike Johnnie, though, Alan really is young and inexperienced, and the novel centers on his sexual awakening in the arms of Evelyn, an event that she forestalls with what the reader can see is mock reticence but that Alan constructs as genuine modesty. Evelyn also presents herself to Alan as a victim of domestic abuse, calculating that this will prompt a protective urge in him. She tells Alan that Fred makes her do "dreadful" things: "I don't know that you could actually call them perversions . . . , but . . . well, they're certainly not very nice."³⁴ But it is also possible to read a degree of complicity between Evelyn and her husband where Alan is concerned, especially when, before a party at their home, Fred dares Alan to try to pass as a girl in front of a guest he has not previously met. This curious and, for Alan, disempowering incident is repeated later when Fred, having discovered the affair, confronts Alan: "Alan's fingers began to play with a heavy paperweight of polished stone that lay near the corner of the desk. It was no more than a reflex action. He did not know that he was doing it. . . . His nerves . . . would no longer obey his intelligence. . . . As instinctively as if he had only spat at him, Alan hurled [the paperweight] at Fred's head."³⁵ He believes he has killed Fred, and Evelyn, apparently horrified, persuades him to dress in women's clothes again in order to make his escape.

In the context of the novel's rudimentary gender politics, dressing as a woman is emasculatory for Alan, and the incident also opens his eyes to Evelyn's self-interest: "He saw now that though Evelyn had loved him well enough in her way, she must always have loved herself better."³⁶ He experiences relief when he is finally able to dress as a man again: "With the feel of a shirt and shorts on him again . . . , Alan had recovered his manhood."³⁷ Even though Alan's gender identification is crudely represented as a matter of clothing, Iles appears to reach toward an exploration of the forces that might influence an individual's self-perception and self-identification. Alan's compliance with Evelyn's demands (and it is perhaps no accident that her name is one that can be used by either a man or a woman) positions him as "feminine" or disempowered; his violent encounter with Fred gives him a taste of what it might mean to occupy a position of hegemonic masculinity.

Whereas in his earlier novels Iles toys with notions of legal as opposed to moral responsibility by constructing murders that might pass as suicides, here the question is how the law might consider a crime that is the result of provocation. The description of Alan's impulsive action is immediately followed by a legalistic statement: "The British law recognizes intolerable provocation to diminish murder into manslaughter. It does not recognize verbal provocation."³⁸ Within the narrative, Alan and Evelyn read a book about a recent famous case, "a wife and a young man, her lover, charged jointly with the murder of the woman's elderly husband," a scenario that contemporary readers would have recognized from the 1935 trial of Alma Rattenbury and her younger lover, George Stoner, for the murder of Alma's husband, Francis, a case that Iles also discussed in an essay.[39] Alan and Evelyn focus on the hypocrisy of the judge and the distorted way that the affair between Rattenbury and Stoner was depicted in the press. Or at least they read into the case the ways in which their own affair might be misunderstood by an outside observer. But any extended parallel with this case, which ended with Alma Rattenbury committing suicide after being acquitted and with Stoner receiving a prison term, is forestalled by the fact that despite his and Evelyn's anxieties, Alan has not killed Fred at all but only injured him. The closing pages of the novel show Alan bouncing back from what initially seems like the traumatic aftermath of the affair and reconstructing it for his friends at university as an exciting adventure in which he is the hero.

The tension between the rational and the instinctive, between control and impulse, is played out even more starkly in *Malice*, especially in the depiction of Bickleigh's sexual adventures. Bickleigh is not restricted to fumbling with his party guests in the potting shed. His work takes him out around the neighborhood by car (and indeed, being out on his rounds provides his alibi for his wife's murder). This gives him a degree of freedom, and before Madeleine Cranmere catches his eye, he is secretly involved with a young woman named Ivy Ridgeway. Their meeting place is a cave that they discover in an old stone quarry:

> Inside, [the cave] was . . . just high enough for the two to be able to stand upright without knocking their heads. They were enraptured with it. The game of cave-dwellers was inaugurated on the spot, and both set to playing it like the children they felt. The cave was furnished, with a half-dressed block of stone for a table, a couple of others for chairs, and armfuls of sweet-smelling bracken for a divan at the farther end. Ivy even went so far as to grow her bobbed hair from that minute, for no cave-dwelling woman is correct without long hair.[40]

This is a rather arid, even comical version of the lovers' pastoral retreat, but it is perhaps intended to serve a similar purpose. For Ivy and Bickleigh, the cave signifies a withdrawal from society and its concerns—a desocialized, primitive form of relationship. But even this retreat cannot be completely sequestered from the wider world. Ivy brings with her gossip from the surrounding community, and later, it is here that she announces that she is pregnant. It is also here that Bickleigh first seriously considers murder. As he begins to tire of

Ivy and is edging his way toward Madeleine, Ivy confronts him with her suspicions and her jealousy, claiming that Madeleine is playing with him:

> "Just amusing herself, and laughing at you behind your back, I expect." ...
> A horrible thought swept into Dr Bickleigh's mind: "Push her over the edge! If you thrust high enough she'll turn over in the air and land with her head on that boulder at the bottom. Nobody could ever know." The longing just for one second was so intense that Dr Bickleigh actually had to clench his hands by his sides to resist it.[41]

Later, when Ivy reveals that she was lying to him about her pregnancy and reinforces her belief that Madeleine is being dishonest with him, a conversation that also takes place at the cave, Bickleigh punches her: "Something took control of Bickleigh. Something that made him hit Ivy in the face with his clenched fist. . . . Dr Bickleigh had never laid any but amorous hands on a woman before. His brain, incapable of thought, was throbbing with mixed emotion: half of it was disgust, and half a queer, shouting exultation."[42] These two incidents, the thought of pushing Ivy over the edge and hitting her, display a violent masculinity that is associated, at least metonymically, with the cave and the desocialized world it represents. This is a space where "primitive," socially untamed feelings can be enacted rather than simply articulated verbally in the form of a wish, as his desire for Julia's death is initially. It is notable, then, that the murder plans that Bickleigh eventually puts into action center not around physical violence but around the abuse of the knowledge of chemistry gained in his medical training. The incidents with Ivy, aside from further eroding any residual sympathy the reader might feel for Bickleigh, expose the latent remnants in him of an impulsive violence that is generally excluded from detective fiction. In detective fiction, the crime must be planned and the evidence concealed so that its gradual revelation can enable the detective to display his intellectual and rational prowess. Opportunism is excluded. But it is also excluded because it is so much harder to explain, let alone rationalize, the kind of impulse that Bickleigh experiences, especially if, as here, it is implied to be a throwback to the primitive.

Conclusion

Iles's desire to shift the focus of the crime narrative to character has consequences not only for characterization, then, but also for narrative form. These novels show him being drawn in different directions. References to historical crimes invite the reader to consider the potential connections between literary and historical criminality, while allusions to forces beyond rationality, however hazily conceptualized, imply that crime will always have a nonrational aspect and further erode the idea of a crime narrative as a purely rational and rationalizable structure. Genre labels are revealed to be not only a flag of convenience for publishers, reviewers, and readers but also a means of attempted control. A "failed" example of a genre may or may not inaugurate a new approach and new form. Iles's troubling and troublesome novels were the most prominent examples of a desire to break new ground in interwar crime writing, and the ways in which they fail are as revealing as their undoubted influence.

Notes

1. Laura Thompson, *Agatha Christie: An English Mystery* (London: Headline, 2008), 366.
2. Malcolm Turnbull, *Elusion Aforethought: The Life and Writing of Anthony Berkeley Cox* (Bowling Green, OH: Bowling Green State University Popular Press, 1996), 79.
3. Ibid., 80, 81.
4. See Anthony Berkeley, "Who Killed Madame 'X'?," and Francis Iles, "Was Crippen a Murderer?," both in Dorothy L. Sayers et al., *Great Unsolved Crimes* (London: Hutchinson, 1938), 112–18 and 15–22. For *Connoisseurs of Crime*, see genome.ch.bbc.co.uk.
5. Tsvetan Todorov, "The Typology of Detective Fiction," in *Modern Criticism and Theory*, ed. David Lodge and Nigel Wood (Harlow, UK: Longman, 2008), 227.
6. Gill Plain, *Twentieth-Century Crime Fiction: Gender, Sexuality and the Body* (Edinburgh: Edinburgh University Press, 2001), 13.
7. Dorothy L. Sayers, introduction to *Great Short Stories of Detection, Mystery and Horror* (London: Victor Gollancz, 1948), 17.
8. Ibid.
9. Dorothy L. Sayers, "Aristotle on Detective Fiction," in *Unpopular Opinions* (London: Victor Gollancz, 1946), 180.
10. Dorothy L. Sayers, "The Present Status of the Mystery Story," *London Mercury* 23 (1930): 51.
11. H. Douglas Thomson, *Masters of Mystery: A Study of the Detective Story* (London: Collins, 1931), 24.
12. Nicola Humble, *The Feminine Middlebrow Novel, 1920s to 1950s* (Oxford: Oxford University Press, 2001), 228.
13. Ibid., 229.
14. Ibid.
15. W. H. Auden, "The Guilty Vicarage," in *The Dyer's Hand* (London: Faber and Faber, 1962), 149.
16. Francis Iles, *Malice Aforethought* (London: Pan, 1979), 7.
17. Ibid., 29–30.
18. Ibid., 29. Bickleigh's relatively small stature as well as some other plot details are borrowed from the case of Herbert Rowse Armstrong, the Hay-on-Wye lawyer who was found guilty of poisoning his wife in 1922.
19. Francis Iles, *Before the Fact* (London: Pan, 1999), 8.
20. Ibid., 34.

21. Ibid., 326.
22. LeRoy Lad Panek, *Watteau's Shepherds: The Detective Novel in Britain, 1914–1940* (Bowling Green, OH: Bowling Green State University Popular Press, 1979), 121.
23. Iles, *Malice*, 73–74.
24. Ibid., 154.
25. See Simon Baatz, *For the Thrill of It: Leopold, Loeb, and the Murder That Shocked Jazz Age Chicago* (New York: Harper Perennial, 2008), esp. 324–25.
26. Patrick Faubert, "The Role and Presence of Authorship in *Suspicion*," in *Hitchcock and Adaptation: On the Page and Screen*, ed. Mark Osteen (Lanham, MD: Rowman and Littlefield, 2014), 44.
27. Iles, *Before the Fact*, 309.
28. Ibid., 43.
29. Ibid., 274.
30. Ibid., 309.
31. Victoria Stewart, *Crime Writing in Interwar Britain: Fact and Fiction in the Golden Age* (Cambridge: Cambridge University Press, 2017), 75–76.
32. Iles, *Before the Fact*, 289.
33. Rick Worland, "Before and After the Fact: Writing and Reading Hitchcock's *Suspicion*," *Cinema Journal* 41, no. 4 (2002): 15.
34. Francis Iles, *As for the Woman* (London: Jarrolds, 1939), 213.
35. Ibid., 267–68.
36. Ibid., 284.
37. Ibid., 313.
38. Ibid., 268.
39. Ibid., 243. See Francis Iles, "The Rattenbury Case," in Dorothy L. Sayers et al., *More Anatomy of Murder* (New York: Berkley, 1990), 59–151.
40. Iles, *Malice*, 44.
41. Ibid., 46.
42. Ibid., 70.

References

Auden, W. H. "The Guilty Vicarage." In *The Dyer's Hand*, 146–58. London: Faber and Faber, 1962.
Baatz, Simon. *For the Thrill of It: Leopold, Loeb, and the Murder That Shocked Jazz Age Chicago*. New York: Harper Perennial, 2008.
Berkeley, Anthony, "Who Killed Madame 'X'?" In Dorothy L. Sayers et al., *Great Unsolved Crimes*, 112–18. London: Hutchinson, 1938.
Faubert, Patrick. "The Role and Presence of Authorship in *Suspicion*." In *Hitchcock and Adaptation: On the Page and Screen*, edited by Mark Osteen, 41–57. Lanham, MD: Rowman and Littlefield, 2014.
Humble, Nicola. *The Feminine Middlebrow Novel, 1920s to 1950s*. Oxford: Oxford University Press, 2001.
Iles, Francis. *As for the Woman*. London: Jarrolds, 1939.
———. *Before the Fact*. London: Pan, 1999.
———. *Malice Aforethought*. London: Pan, 1979.
———. "The Rattenbury Case." In Dorothy L. Sayers et al., *More Anatomy of Murder*, 59–151. New York: Berkley, 1990.
———. "Was Crippen a Murderer?" In Dorothy L. Sayers et al., *Great Unsolved Crimes*, 15–22. London: Hutchinson, 1938.
Panek, LeRoy Lad. *Watteau's Shepherds: The Detective Novel in Britain, 1914–1940*. Bowling Green, OH: Bowling Green State University Popular Press, 1979.
Plain, Gill. *Twentieth-Century Crime Fiction: Gender, Sexuality and the Body*. Edinburgh: Edinburgh University Press, 2001.
Sayers, Dorothy L. "Aristotle on Detective Fiction." In *Unpopular Opinions*, 178–90. London: Victor Gollancz, 1946.
———. Introduction to *Great Short Stories of Detection, Mystery and Horror*,

edited by Dorothy L. Sayers, 11–26. London: Victor Gollancz, 1948.

———. "The Present Status of the Mystery Story." *London Mercury* 23 (1930): 47–52.

Stewart, Victoria. *Crime Writing in Interwar Britain: Fact and Fiction in the Golden Age.* Cambridge: Cambridge University Press, 2017.

Thomson, H. Douglas. *Masters of Mystery: A Study of the Detective Story.* London: Collins, 1931.

Thompson, Laura. *Agatha Christie: An English Mystery.* London: Headline, 2008.

Todorov, Tsvetan. "The Typology of Detective Fiction." In *Modern Criticism and Theory*, edited by David Lodge and Nigel Wood, 226–32. Harlow, UK: Longman, 2008.

Turnbull, Malcolm. *Elusion Aforethought: The Life and Writing of Anthony Berkeley Cox.* Bowling Green, OH: Bowling Green State University Popular Press, 1996.

Worland, Rick. "Before and After the Fact: Writing and Reading Hitchcock's *Suspicion.*" *Cinema Journal* 41, no. 4 (2002): 3–26.

Part 2
The Body Captured

5.

The Art of Identification
The Skeleton and Human Identity

Rebecca Gowland
and Tim Thompson

Establishing identity from skeletal remains is a fundamental part of the practice of bioarchaeologists and forensic anthropologists. It is the job of the former to analyze skeletons of the long dead and of the latter to analyze those of the recently deceased, to record as much information as possible about these people and to interpret this evidence in terms of how they lived and died. Rebecca is a bioarchaeologist, focusing on the analysis and interpretation of the skeletal remains of individuals excavated from archaeological contexts. Bioarchaeologists are adept at recording a range of macroscopic and biomolecular indicators of past diet, disease, demography, mobility, social identity, interactions, and subsistence. Tim is a forensic anthropologist whose remit generally concerns the identification and interpretation of the remains of individuals from contexts pertinent to a court of law (usually within the past seventy years). Forensic contexts may vary widely in scope, including mass fatality incidents (e.g., plane crashes or natural disasters, such as the Southeast Asian tsunami of 2004), areas of conflict (e.g., the exhumation of mass graves), or individual deaths through accident or foul play. It is important to note that the techniques used to analyze and interpret skeletal remains are identical in both forensic and archaeological settings. Differences between the disciplines relate largely to contextual distinctions as well as variation in the types of questions asked of the data.

Traditionally, forensic anthropology as a discipline has been less engaged with social theory than have other academic domains; it aligns itself more closely with biomedical and clinical models, which have tended toward the universality of the human body. Bioarchaeology, by contrast, has moved toward a much more socially integrated approach to the skeletal evidence, particularly over the past decades.[1] This divergence between the two disciplines stems partly from the difference in the questions posed and partly from the different disciplinary backgrounds of forensic and bioarchaeological practitioners (particularly when comparing the United Kingdom and the United States). Bioarchaeologists are interested in recording human skeletal remains, not in a purely biomedical manner but in terms of what they can tell us about how people lived in the past—the emphasis tends to be on the broader context rather than on the individual. By contrast, the forensic practitioner is often focused on documenting individualizing features of the skeleton and on factors that can help establish individual identity. Operating largely within a medico-legal paradigm, the forensic scientist has a tendency to conceptualize the skeleton as an inert universal entity: in forensics, "human identification" is considered to be an objective scientific endeavor, with the body conceptualized in predominantly biological terms. Within this framework, variation in skeletal features as a consequence of lifestyle or other social factors is conceptualized as unwanted variation or "noise" that affects the efficacy of the techniques.

The aim of this chapter is twofold. First, we discuss the social life of the skeleton and how we can access different aspects of social identity from skeletal remains. Second, we undertake a critical review of the scientific model and the methods used to analyze the skeleton in forensic and bioarchaeological contexts. The human skeleton is a dynamic living tissue that interacts with and is affected by the social as well as physical environment in which someone lives.[2] The body is never a purely biological entity because it is molded and altered by social experiences, from the biomolecular to the macroscopic level, throughout life. Neither is skeletal analysis an entirely objective, "pure" science, because we imbue the skeleton with meaning that is culturally situated, and our methods have to be reflexive and versatile in order to accommodate the range of human variation we encounter. This chapter examines some of these factors, providing examples of scientific dilemmas in determining identity within forensic and bioarchaeological contexts, including those in which biomolecular and macroscopic techniques are used. It will discuss sex determination, DNA analysis, epigenetic developments, and recent critiques exploring cognitive bias within the disciplines.

The Body and Society

Until the 1990s the body was largely absent from the social science literature, relegated to the status of a biological vessel in which the human social agent happened to reside.[3] Disciplinary divides operating along the science/social theory and mind/body models meant that the biological aspects of the body became the preserve of those engaged with scientific discourses, while social identity as a culturally specific and historical construct was a distinctive field of study in which the physical body was viewed as a largely passive "absent presence."[4] With the pioneering work of sociologists such as Bryan S. Turner and Chris Shilling, the role of the physical body in social identity and the dialectical relationship between the two started to become a new focus of study.[5] Flesh, blood, and bones for the first time started to feature in the social science research agenda. In archaeology, the science/social theory divide was and still is somewhat entrenched within the discipline. Social constructionism had been highly influential for examining the fluidity and intersectionality of identity in the past, but many of these approaches marginalized the physical body. The concepts of embodiment and the embodiment of identity emerged as a prominent theme in archaeology during the late 1990s and 2000s because of work by influential authors such as Lynn Meskell and Rosemary Joyce.[6] Archaeological discourse started to engage with the body as a physical entity rather than as simply a passive clotheshorse for material culture.[7] The importance of the physical remains of the body for reconstructing past lifeways has been a focus of study in bioarchaeology for many years. The explicit theorization of these remains, however, as the physiological embodiment of social processes and integration with social theory has surfaced only more recently.[8]

The forensic context has been much slower to acknowledge and engage in this complex discussion. The strong emphasis on the application of science as an objective truth has meant that such discourse has been discouraged or marginalized. Forensic anthropology in the United Kingdom emphasized the *forensic* aspect; thus with the development of the subject and the increasing prominence of British forensic anthropology in the academic literature in the early 2000s came a consolidation of the view that forensic anthropology was an atheoretical exercise in following skeletal evidence to the "truth." As American forensic anthropologists (particularly those influenced by Latin

American teams) gained an increasing global visibility, so came a greater emphasis on the *anthropology* aspect (a result of the entrenched disciplinary distinctions on either side of the Atlantic). Today, a much greater interest is emerging in the sociocultural anthropological perspective on the subject, and with this comes tensions in the discipline, particularly regarding the internationalization of forensic anthropological practice and the thorny issue of standardization.[9] In this chapter we will provide both macroscopic and biomolecular examples to illustrate the interaction between social identity and the hard tissues of the body, and we will offer a critical review of the art of human identification within our respective disciplines.

Human Identification in Forensic Anthropology and Bioarchaeology

To the untrained eye, skeletons are all the same: the bones and teeth are inert tissues that are more or less identical among individuals. To the forensic anthropologist/bioarchaeologist, the human skeleton is a tangle of variation. Bodies are messy and unruly; they confound scientific scrutiny and elide neat categorizations. Bodies do not subscribe to linear, defined, discrete trajectories and subdivisions, and their dimensions and morphology do not readily lend themselves to systematic, universally applicable methods of analysis. This is both a blessing and a curse but is more often viewed as the latter; such transgressions from the type exacerbate error and undermine conclusions. Nevertheless, we conduct our practices, applying our methods (including appropriate error ranges) with the aim of "reading" information from the bones in a clinically objective and standardized way. We adopt the techniques of scientific integrity and objectivity, but there exists an underlying anxiety about the problematic nature of our methods and our categorizations. This unease stems partly from the fact that the reference collections on which our osteological methods are devised may bear only a passing resemblance to our target samples in terms of detailed biological parameters. This is because skeletal features are contingent on the interplay between genes and environment, and this is challenging, if not impossible, to mitigate. We know that our error ranges, therefore, may represent only a loose approximation because our methods are confounded by the enormity of variation in the human skeletal phenotype.

This situation creates a mismatch between the required forensic deliverables, the expectation of "hard facts" expounded by the popular media on TV programs such as *Bones* or *CSI*, and the realities of practice in which our "certainties" become submerged by the specificity of circumstance. Let us take an apparently straightforward question such as this: how long does it take the body to decompose in the ground? The police would wish us to give an answer along the lines of, for example, five years and two months, and as scientists we certainly strive to deliver such precision. Unfortunately, decomposition of the human body is contingent on a range of factors intrinsic to each person, such as age at death, sex, health status, body mass, and the presence of trauma. Extrinsic factors relating to the mode of deposition (which we call taphonomic factors)—such as whether the body was buried, whether the body was wrapped or contained, grave depth, soil type, rainfall, temperature, and local flora and fauna—create further variability. These variables can be accounted for only to a certain extent because they interact in multiple ways that are difficult to model using experimental science.

Similarly, the question of how old a person was when they died would appear to be straightforward; after all, we know that the body grows and develops, reaches maturity, and then gradually senesces according to a broad chronological schedule. Except that unfortunately there is enormous variability in the rate of maturity and senescence that just increases with age. Although infants and children may hit biological milestones according to a general timetable, the interactions between genetic and environmental factors also create a significant amount of variation. Factors such as intrauterine environment, gestational age and size at birth, infant feeding practices, childhood diseases, social environment, and genetic growth potential will all have an impact on the relationship between biological and chronological age. The situation becomes far more problematic when we examine the adult skeleton, as skeletal changes associated with age are also very subtle and can lack conformity with the "type" features of established methods. Indeed, the process of estimating age at death from the adult skeleton was long ago described as an "art rather than a precise science."[10]

All this emphasis on subjectivity and variability may be deemed vexing and even appear to undermine the discipline; it sheds doubt on the scientific rigor of carefully constructed methods. It presents too many shades of gray in a context (i.e., court of law) that insists on black and white and the unassailability of facts. It flies in the face of the Daubert agenda, which dictates the criteria that every forensic discipline must adhere to in order to be accepted by the US courts. The intention of this chapter is not to undermine the great work that is ongoing within the discipline. It does not aim to detract from the cumulative scientific developments of the past 150 years that seek to refine, improve, and standardize methods. Indeed, much of our own research has strived to contribute to this agenda. We argue instead that there is no need to downplay the uncertainties that plague our discipline, because they are also our strength. The body is a social project and therefore subject to intense variability; we are born into and grow up in social worlds, which gradually become enmeshed in the hard tissues of our bodies, altering them in incremental and sometimes profound ways. We are not preprogrammed biological bots that (barring illness or accident) blunder through the predetermined genetic blueprint of our lives. We know that our diet, place of origin, activities, social interactions, identities, and relationships all conspire to create phenotypic diversity. As we discuss later, epigenetic research shows how genes can be switched on or off depending on social and environmental adversity. The body is never just a blank slate on which cultural practices are gradually inscribed: the body is and always has been a cultural entity.

In the following text we highlight a number of issues. First, the identification categories we use may at times be problematic, and this is partly because they are an artificial construct. Categories are our means of ordering the natural world, but with human skeletal analysis, in which very minor morphological variants become important, exceptions to the rule are more common than we may care to acknowledge. We do not wish to sound too negative here: our methods for ordering and interpreting human bodies have a great deal of proven success. We know that we can approximate age at death in most cases and that we can quite accurately determine sex. But not in all instances. And that's okay. It is not a weakness of our work; it is a strength to acknowledge the reality of our bodies. We should be secure enough in our practice to own and highlight such difficulties, which after all stem not from our own deficiencies as scientists but from the fact that humans and their bodies are variable, contradictory, capricious, and inconstant. At times, we can harness this variability to arrive at individualizing features that help establish a person's identity, but at other times this variability must be accounted for statistically so that, for example, we do not provide misleading information to the police.

Gender Jeopardy

The distinction between sex as a biological variant and gender as the cultural interpretation of biological difference has been extremely influential within the social sciences. It has been important for facilitating a more critical perspective on masculinity and femininity, not as ahistorical biological and dualistic frameworks but as fluid cultural constructions.[11] Feminist scholars have noted since the 1990s, however, that an unintended consequence of this distinction between sex and gender has been that the natural, ahistorical condition of the human body with respect to sex has been largely unchallenged.[12] Cultural constructions of gender identity are perceived to overlay the body but not touch or modify it in any tangible sense. The sex/gender divide also reflects disciplinary subdivisions such that biological sex is the concern (in this context) of bioarchaeologists/forensic anthropologists, while gender is a subject for cultural anthropologists/social archaeologists. Numerous scholars have challenged the sex/gender distinction, casting particular doubt on the scientific immutability of biological sex.[13] The bioarchaeological literature recognizes the problematic nature of the divide but also argues that there is utility in maintaining this differentiation.[14] Forensic scientists, like clinicians, tend to use the terms *sex* and *gender* interchangeably and until recently have not engaged with the theoretical aspects of these debates. For forensic anthropologists, gender as a cultural concept falls beyond their remit: culture may increase the noise of biological variability, so it should be accounted for but not theorized. Gendered artifacts such as clothing that may be recovered with the body are considered to be problematic for identification because they are not stable and can thus be moved from one person to another, potentially confounding identification. The biological body, by contrast, is viewed as stable and unchanging. Within bioarchaeology, there has been a much more explicit recognition of the importance of body/society interactions in which gendered ideologies may imprint and physically alter the human body.[15]

The skeleton is never fully biological in terms of sex due to how society constructs gendered difference and the impact that this has on bone physiology. As Anne Fausto-Sterling notes, "One cannot easily separate bone biology from the experience of individuals growing, living and dying in particular cultures and historical periods and under different regimens of social gender."[16] For example, male and female infants may receive different treatment in terms of care and diet from birth onward as the result of perceived cultural value, thus exposing infants to differential risks of nutritional deficiencies and infectious diseases that can affect growth, stature, morbidity, and mortality in later life. Several authors have also described how the skeleton becomes gendered and modified by the culturally constructed roles ascribed to different sexes. These modifications can be expressed, for example, in differences between males and females in patterns of joint disease, skeletal indicators of activity, and traumatic injury.[17]

Some have suggested that the scientific terminology currently used to describe male and female skeletal features feeds into contemporary views about masculinity and femininity and may even impact scientific practice. For example, features of the male skull and pelvis are described as "rugged," "robust," and "heavy," while females are described as "delicate," "gracile," and "fine." This use of language creates a polarization of skeletal sexual characteristics that belies the considerable overlap between them. For example, Walker's study of the sexually dimorphic characteristics of the skull highlighted both the degree of variation between male and female traits in skeletons from different populations and the degree of overlap in features between males and females.[18]

Sexually dimorphic skeletal traits do allow scientists to identify male and female skeletons with strong levels of success. It is important to acknowledge, however, that these traits exist along a spectrum from hyperfeminine to hypermasculine, as demonstrated in our scoring of traits as Male, Possible Male, Intermediate, Possible Female, and Female. It is not uncommon to encounter skeletons that exhibit features that are intermediate or even a mosaic of traits (i.e., a mixture of masculine and feminine characteristics), particularly with respect to the skull. Forensic anthropologists and bioarchaeologists are aware that sexually dimorphic features are scalar and that parameters can shift substantially between skeletal populations. This occurs to the extent that a particular trait may fall within the female range in one population sample and closer to the male range in another. This is an interesting fact that surprises people outside of the discipline and yet receives remarkably little commentary by practitioners who use standardized terminology and "scoring systems" to assign sex to skeletal remains. There is no formal method of accounting for population differences in sexual dimorphism; instead, we make mental adjustments to the established criteria once we have gauged the nature and extent of sexual difference within a particular sample. These adjustments are rarely noted or formalized; they simply exist within the cognitive realm of the scientist recording the skeletons. Forensic anthropologists and bioarchaeologists know and accept that this must happen. We cannot be dogmatic and unyielding in our application of methods in the face of profound human variation; to do so would hinder accuracy. Yet this aspect of practice is rarely documented—it is as though there is a fear that in doing so our methods would appear less scientifically objective and therefore less valid.

It is not actually possible to create a formal protocol for such ad hoc adjustments, however, as they are population specific and predicated on the experience of the practitioner. We should be explicit about the fact that we do this, however; it demonstrates that we acknowledge the interplay of biology and culture in the formation of sexually dimorphic features; that we embrace variation and harness it as a tool for understanding past and present populations; and that we recognize that sexual dimorphism exists on a spectrum that resists the polarizing discourse of male-/female-only categories.

It has been shown that while dimorphic characterizations are usually possible, for every physiological measure of human sexual difference—from DNA, hormones, and genitalia to bones—there is a spectrum of states.[19] Other cultures are more comfortable with respect to the fluidity of both gender and biological difference in sexual characteristics. One high-profile area in which the problematic nature of our ideas about biological sex is evident is in elite-level female athletics. Several female athletes in recent years have been subjected to humiliating rounds of gender testing to establish whether they are definitively a man or a woman. In the high-stakes world of athletics this is a crucial question, given the biological advantage of male traits in many track-and-field events. Hormone testing and genetics have been the tools of choice for "outing" these women as "masculine" and therefore as cheaters. What such tests and scandals have revealed is the ambiguity of this seemingly straightforward question and the disconnect between contemporary dualisms relating to sex and the biology of the human body.[20] By acknowledging the spectrum of phenotypic sexual variation within our own practice we are not undermining our techniques but instead opening up our analysis to include the spectrum of human embodied experiences in both the past and the present.

Contrary to what you may have heard, DNA is not the answer to every forensic question. When Sir Alec Jeffreys highlighted that our DNA profiles were unique and could be used for individualization, the potential of DNA analysis for the forensic sciences increased exponentially. Indeed, it is hard to imagine a more significant development in the forensic arena. The underlying biology is deceptively simple—each of us (except for identical twins) has a unique DNA code essentially comprising a long sequence of just four bases: thymine (T), adenine (A), cytosine (C), and guanine (G). The combination of these four parts over such a long strand is unique. Forensic methods of analyzing DNA samples focus on a few restricted areas of this long chain, but even this has enough discriminatory power to separate and identify individuals. It is worth noting here that forensic profiling generally uses nuclear DNA, which can be found in most cells. There is another type of DNA, mitochondrial DNA, which is more prevalent in the body but is derived only from the mother and thus can determine only maternal lineage. The subsequent approaches used to support decision-making about whether a DNA sample matches a given person are highly statistical, giving the impression that the method is quantitative, immutable, and incontrovertible. Within the public consciousness, when compared to all other strands of forensic science, DNA profiling occupies the most elevated position when it comes to providing the ultimate evidential truth. Likewise, within bioarchaeological research, DNA analysis has yielded startling new insights into human evolution; the origin and dispersal of peoples, animals, and diseases; and a whole host of transformational changes. Studies involving DNA analysis tend to be published in high-profile, high-impact journals and to attract large amounts of funding. Yet, ancient DNA evidence can also be problematic, with grand interpretations often made based on very small samples. Issues of contamination and poor preservation can be rife, and there is a tendency for more traditional forms of archaeological evidence to be disregarded in the face of DNA because of the position that DNA studies hold in the scientific evidential hierarchy.

Unfortunately, the application of DNA profiling in forensic anthropology is not as straightforward as it seems. There are several issues to contend with, and as with bioarchaeology, the most significant is contamination. Since DNA profiling uses such small fragments of the human body, it works only when you can be certain that the tiny sample you are analyzing is from the person of interest. Contamination by other deceased victims, by CSIs, or by forensic investigators will scupper the conclusions. The famous white crime-scene suits that CSIs and others wear are a means to mitigate this. But research has shown that such contamination can be subtle and unexpected—even bones can show evidence of contamination from surrounding decomposing remains. Furthermore, what the public often does not immediately understand is that identification of an individual based on DNA profiling requires a DNA database or an available sample from a close relative to compare the bodily sample to. Without one of these, the identification process will fail.

The final issue to reflect on is the *CSI effect*. This term has been adopted by the public, the media, and forensic science to reflect the possible impact that TV crime shows have on the perception of forensic methods. The upshot is that the public assumes that DNA profiling is quicker, easier, and more reliable than it really is—or so it is thought. While the concept is a source of great concern within the forensic sciences, some argue that the actual evidence for this impact is negligible and that the effect is just an artificial creation.[21] Nonetheless, discussion of the CSI effect has worked its way into forensic anthropology. For example, investigations into the Spanish Civil War graves relied heavily on DNA profiling to identify the deceased. The public felt that this approach would result in a speedy resolution of the identification issues, but the complexity of the graves, the possibility of contamination, and the challenges of simply recovering DNA from bone meant that the process was anything but straightforward.[22] The CSI effect has frequently led to false expectations of what is achievable with DNA, regardless of circumstances, and in the case of the Spanish Civil War graves, this effect has impacted subsequent attempts to work with the remains because public confidence in DNA profiling was shaken.

Cognitive Conundrums

We all know that our mind can play tricks on us. Unfortunately, this is problematic when working in a forensic context in which the outcome of scientific investigations can have very real life-or-death implications. One consequence of regarding forensic anthropology in particular as an objective science is that there has been little interest in exploring the human factors involved in the application of methods and the interpretation of results. As was noted earlier, human variation is a constant consideration when examining human remains, but it is also a factor intrinsic to the scientific observer. Decision-making is integral to every step of the osteological process—from choosing which methods to use to internally scaling for sexual dimorphism to determining the conclusions of an age estimation method. Bioarchaeology and forensic anthropology are experiential disciplines, yet we understand little about the impact of those experiences. We do know, though, that they have some influence. Specifically, we know that contextual information presented alongside human remains can have a significant influence on the outcome of decision-making. For example, it has been shown that contextual information often results in confirmation bias when conclusions are made following osteological analysis.[23] Further work by Sherry Nakhaeizadeh and colleagues has also demonstrated that the diagnosis of traumatic injury to the skeleton was influenced by the accompanying text (i.e., multiple individuals presented with the same injury interpreted it differently solely because of the text that accompanied it).[24] In all cases the practitioner's level of experience was shown to be an important variable in how much influence the accompanying text had.

This clearly presents a challenge to forensic anthropology. How can we offer the irrefutable conclusions demanded by the courts if our conclusions differ because of our own biases and personal and professional experiences and because of varying contextual information? This is an issue that has vexed the leading cognitive neuroscientist Itiel Dror for many years, and he provides some solutions in a recent paper in *Science*:

> Biases . . . can be minimized by case managers who ensure that only relevant information gets to the appropriate expert. . . . Bias cascade and bias snowball can be minimized by compartmentalization. For example, the person collecting evidence from a crime scene should not be the expert who analyzes that data in the laboratory. In that way, any exposure to extraneous information at a crime scene does not influence the subsequent analysis.[25]

This solution suggests that this is a straightforward data-management issue that can be further minimized when combined with standardized methods of analysis. This does not, however, account for how the differing experiences of practitioners influence decision-making or for the judgment of some information as relevant and some as irrelevant. As we have taken great pains to highlight, it is the marriage of biology and social context that gives the human skeleton its power. It is important that practitioners be reflexive in their work while still operating within established scientific parameters.

Epigenetic Epiphanies

Molecular epigenetics is a rapidly developing field of research that is challenging our understanding of the body/society paradigm. Research in this field highlights the role of environmental stimuli, including social forces, in altering patterns of gene expression (not the genome itself) that result in changes to biological responses. These epigenetic alterations include the propensity to develop chronic physical and mental disease as a consequence of adversity experienced during fetal and infant life. Epigenetic mechanisms are sometimes described as a "molecular memory of past stimuli."[26] These alterations may even result from adversity experienced by immediate ancestors, which creates epigenetic changes that are transmitted intergenerationally.[27] A well-known example of this impact has been evidenced via research on victims of the Dutch Hunger Winter during World War II—a five-month period of famine brought about through Nazi food blockades. Longitudinal studies of individuals who endured this famine *prenatally* (and their children in turn) were shown to have increased risks of metabolic disorders such as diabetes and cardiovascular disease as adults.[28] It is not just nutrition that influences gene expression; factors such as maternal stress level and the quality of caregiving during infancy are also significant for later mental and physiological well-being.[29] For example, differences in maternal care of infants have long-lasting consequences for epigenetic processes that help regulate adult stress reactivity.[30]

The significance of these intergenerational epigenetic effects for the interpretation of bioarchaeological and forensic evidence is potentially profound.[31] Epigenetic processes show us that our life histories are essentially intertwined with our ancestors and our descendants; we do not have discrete, bounded biographies with a clear beginning and end. This presents an important biographical challenge to current life-course research: social factors affecting our grandmothers have repercussions for our own ontogeny and embodied identity.[32] It also highlights issues with the interpretation of some bioarchaeological data. We often interpret pathological lesions on the skeleton as due to immediate causalities; for example, growth stunting in young children results from poor diet. The epigenetic model, however, also implicates the mother's and even grandmother's health status when considering growth trajectories.

A variety of studies have highlighted the ways in which social circumstances such as inequality (as a consequence of status, gender, and ethnicity) can produce epigenetic alterations.[33] Through this mechanism, aspects of society come to be literally embodied within the biological tissues and serve to exacerbate health and structural inequalities.[34] While problematic to isolate, epigenetic processes will have observable skeletal outcomes; they are an embedded part of the process of growth and development. Epigenetic research is a move away from the genetic determinism of the 1990s in which our fortunes were conceptualized as being determined by a fixed genome, because it shows us that even at a genetic level we are malleable and influenced by social environment. Some medical anthropologists have argued that there is a danger that individuals will be reduced instead to their epigenome, for example in health-care treatment, and that this would represent a return to the deterministic molecularization of social processes.[35] We would argue instead that this research represents an exciting new avenue for investigation and for capturing the impact of social processes on the body. As Maurizio Meloni states, "Epigenetic research illustrates exemplarily how we are moving toward a post-dichotomous view of biosocial processes." He also says that "the social assumes a causative role in human biology as never before."[36]

Conclusions

This chapter has been wide-ranging but has attempted to provide some snapshots of the variability of the human body and the problems that forensic anthropologists and bioarchaeologists face when capturing and transcribing this variation in a meaningful way. Science strives for objectivity and replicability, but scientists do not operate within a social vacuum—a factor that has been explored by many scholars from a range of disciplines. The methods we use are produced within particular cultural understandings of identity, as the earlier example of sex determination showed; even though skeletons express a spectrum of masculine and feminine traits, we are compelled to conform to the dualistic male/female categorization to derive meaning that fits with our Western worldview. Furthermore, studies of cognitive bias have shown that practitioners are impacted by the social context in which they work and are affected by a range of contextual factors. This is not to say that skeletal remains should be interpreted without recourse to contextual information; indeed, an understanding of context is considered essential in bioarchaeology because, for example, the significance of different pathologies varies according to the periods and places in which they occur.

Our methods of skeletal analysis also have flaws that are rooted in human variability. These methods (e.g., the skeletal age estimation of adults) often rely on extremely subtle morphological differences in skeletal features, and thus any variation/deviation between or within skeletal samples disproportionately affects the results. To mitigate this, we are advised to use methods derived from genetically and culturally similar skeletons with known identities. In practice this is not possible, because such known skeletal collections are relatively rare. Our only other means of mitigating this variation is to account for it statistically by increasing error ranges. But again, we unfortunately cannot quantify such variation in any straightforward way because we have no way of knowing the extent of the phenotypic differences arising from the lack of genetic and cultural parity between the reference skeletal sample and the target sample. This is particularly problematic in the forensic arena and for the implementation in the United States of the Daubert criteria, which consider statistical knowns paramount for scientific evidence in the courts of law.

The diverging aims of bioarchaeology and forensic anthropology have led to differences in approaches and philosophies during recent decades, but there should be greater communication and integration between the two fields, as there is still much common ground. There is also some discord *within* both disciplines, particularly around the different methods used and standards adopted in different countries. The United Kingdom is more closely allied with the United States in scientific approaches and standards, but these may differ from continental Europe. Language barriers create significant scientific divides in the recording and interpretation of skeletal evidence.

Anthropology and archaeology are magpie disciplines; they are outward looking and keen to integrate theories, methods, and research from the broader social and medical sciences as well as the arts and humanities. These qualities are two of our strengths. But subdisciplines in the pursuit of specific goals can become introspective and/or, in an attempt to be taken seriously within medico-legal contexts, strive too hard to deliver black-and-white facts and downplay the many shades of gray. It is considered pejorative to call human identification an art rather than a science. We would argue that this does not demean the discipline but instead elevates it as well as the scientists engaged in skeletal analysis. It recognizes the complexity of the skills required, the need for a nuanced and reflexive application of methods, and the importance of experience. Human variability, in our view, should be exposed and laid bare; the dynamic nature of the body's hard tissues and the cultural specificity of both subject and observer create a challenging but vibrant academic arena. Our cognate disciplines are responsible for and have instituted multiple scientific organizations, annual meetings, and specialist journals; we are certain that we are secure enough as scientists/academics/practitioners to embrace and further explore this variation rather than to obfuscate and conceal it for fear of being undermined.

Notes

1. See, for example, Rebecca Gowland and Christopher Knüsel, *Social Archaeology of Funerary Remains* (Oxford, UK: Oxbow Books, 2006); Joanna R. Sofaer, *The Body as Material Culture: A Theoretical Osteoarchaeology*, vol. 4 (Cambridge: Cambridge University Press, 2006); and Sabrina C. Agarwal and Bonnie A. Glencross, *Social Bioarchaeology* (Chichester, UK: Wiley-Blackwell, 2011).

2. Gowland and Knüsel, *Social Archaeology of Funerary Remains*.

3. Chris Shilling, *The Body and Social Theory* (London: Sage, 2012).

4. Ibid.

5. Bryan S. Turner, *The Body and Society: Explorations in Social Theory* (Oxford, UK: Blackwell, 1984); Shilling, *Body and Social Theory*.

6. Lynn Meskell and Robert W. Preucel, *Companion to Social Archaeology* (Malden, MA: Blackwell, 2004); Rosemary A. Joyce, "Archaeology of the Body," *Annual Review of Anthropology* 34 (2005): 139–58.

7. Gowland and Knüsel, *Social Archaeology of Funerary Remains*.

8. Rebecca Gowland and Tim Thompson, *Human Identity and Identification* (Cambridge: Cambridge University Press, 2013).

9. Zoe Crossland, "Evidential Regimes of Forensic Archaeology," *Annual Review of Anthropology* 42 (2013): 121–37; Hugh Tuller, "Identification Versus Prosecution: Is It That Simple, and Where Should the Archaeologist Stand?," in *Disturbing Bodies: Perspectives in Forensic Anthropology*, ed. Zoe Crossland and Rosemary A. Joyce (Santa Fe, NM: School for Advanced Research Press, 2015), 85–102; Tim Thompson et al., "Forensic Anthropology: Whose Rules Are We Playing By?—Contextualizing the Role of Forensic Protocols in Human Rights Investigations," in *War Crimes Trials and Investigations* (London: Palgrave Macmillan, 2018), 59–80.

10. William R. Maples, "The Practical Application of Age-Estimation Techniques," in *Age Markers in the Human Skeleton*, ed. Mehmet Yaşar İşcan (Springfield, IL: Charles C. Thomas, 1989), 323.

11. Shelley Budgeon, "Identity as an Embodied Event," *Body and Society* 9, no. 1 (2003): 35–55; Pamela L. Geller, "Conceiving Sex: Fomenting a Feminist Bioarchaeology," *Journal of Social Archaeology* 8, no. 1 (2008): 113–38; Pamela L. Geller, *The Bioarchaeology of Socio-Sexual Lives: Queering Common Sense About Sex, Gender, and Sexuality*, (Zug, CH: Springer International, 2016).

12. Nelly Oudshoorn, *Beyond the Natural Body: An Archaeology of Sex Hormones* (London: Routledge, 1994).

13. Judith Butler, *Bodies That Matter: On the Discursive Limits of Sex* (London: Routledge, 1993); Thomas Laqueur, *Making Sex: Body and Gender from the Greeks to Freud* (Cambridge, MA: Harvard University Press, 1990).

14. Marie Louise Stig Sørensen, *Gender Archaeology* (Hoboken, NJ: Wiley, 2000).

15. Sofaer, *Body as Material Culture*; Sandra E. Hollimon, "Sex and Gender in Bioarchaeological Research: Theory, Method, and Interpretation," *Social Bioarchaeology*, 2011, 147–82.

16. Anne Fausto-Sterling, "The Bare Bones of Sex: Part 1—Sex and Gender," *Signs: Journal of Women in Culture and Society* 30, no. 2 (2005): 1510.

17. Joanna R. Sofaer Derevenski, "Sex Differences in Activity-Related Osseous Change in the Spine and the Gendered Division of Labor at Ensay and Wharram Percy, UK," *American Journal of Physical Anthropology* 111, no. 3 (2000): 333–54; Hollimon, "Sex and Gender," 147–82.

18. Phillip L. Walker, "Sexing Skulls Using Discriminant Function Analysis of Visually Assessed Traits," *American Journal of Physical Anthropology* 136, no. 1 (2008): 39–50.

19. Oudshoorn, *Beyond the Natural Body*; Katrina Karkazis et al., "Out of Bounds? A Critique of the New Policies on Hyperandrogenism in Elite Female Athletes," *American Journal of Bioethics* 12, no. 7 (2012): 3–16.

20. Karkazis et al., "Out of Bounds?," 3–16.

21. Simon Cole and Rachel Dioso-Villa, "Investigating the CSI Effect: Media and Litigation Crisis in Criminal Law," *Stanford Law Review* 61 (2010): 1335; Simon Cole, "A Surfeit of Science: The 'CSI Effect' and the Media Appropriation of the Public Understanding of Science," *Public Understanding of Science* 24, no. 2 (2015): 130–46.

22. Francisco Ferrándiz, "Exhuming the Defeated: Civil War Mass Graves in 21st-Century Spain," *American Ethnologist* 40, no. 1 (2013): 38–54; Gillian Fowler and Tim Thompson, "A Mere Technical Exercise? Challenges and Technological Solutions to the Identification of Individuals in Mass Grave Scenarios in the Modern Context," in *Human Remains and Identification: Mass Violence, Genocide and the "Forensic Turn*," ed. Élisabeth Anstett and Jean-Marc Dreyfus (Manchester: Manchester University Press, 2015), 117–41; Thompson et al., "Whose Rules?," 59–80.

23. Sherry Nakhaeizadeh, Itiel E. Dror, and Ruth M. Morgan, "Cognitive Bias in Forensic Anthropology: Visual Assessment of Skeletal Remains Is Susceptible to Confirmation Bias," *Science & Justice* 54, no. 3 (2014): 208–14; Sherry Nakhaeizadeh et al., "Cascading Bias of Initial Exposure to Information at the Crime Scene to the Subsequent Evaluation of Skeletal Remains," *Journal of Forensic Sciences* 63, no. 2 (2018): 403–11.

24. Sherry Nakhaeizadeh, Ian Hanson, and Nathalie Dozzi, "The Power of Contextual Effects in Forensic Anthropology: A Study of Biasability in the Visual Interpretations of Trauma Analysis on Skeletal Remains," *Journal of Forensic Sciences* 59, no. 5 (2014): 1177–83.

25. Itiel E. Dror, "Biases in Forensic Experts," *Science* 360, no. 6386 (20 April 2018): 243.

26. Hannah Landecker and Aaron Panofsky, "From Social Structure to Gene Regulation, and Back: A Critical Introduction to Environmental Epigenetics for Sociology," *Annual Review of Sociology* 39 (2013): 336.

27. Christopher Kuzawa, "Developmental Origins of Life History: Growth, Productivity, and Reproduction," *American Journal of Human Biology* 19, no. 5 (2007): 654–61; Hannah Landecker, "Food as Exposure: Nutritional Epigenetics and the New Metabolism," *BioSocieties* 6, no. 2 (2011): 167–94.

28. Felicia M. Low, Peter D. Gluckman, and Mark A. Hanson, "Developmental Plasticity, Epigenetics and Human Health," *Evolutionary Biology* 39, no. 4 (2012): 650–65.

29. Clyde Hertzman, "Putting the Concept of Biological Embedding in Historical Perspective," supplement, *Proceedings of*

the *National Academy of Sciences* 109, no. S2 (16 October 2012): 17160–167.

30. Zaneta M. Thayer and Christopher W. Kuzawa, "Biological Memories of Past Environments: Epigenetic Pathways to Health Disparities," *Epigenetics* 6, no. 7 (2011): 798–803; Maurizio Meloni, "The Social Brain Meets the Reactive Genome: Neuroscience, Epigenetics and the New Social Biology," *Frontiers in Human Neuroscience* 8 (2014): 309.

31. Rebecca Gowland, "Entangled Lives: Implications of the Developmental Origins of Health and Disease Hypothesis for Bioarchaeology and the Life Course," *American Journal of Physical Anthropology* 158, no. 4 (2015): 530–40.

32. Gowland, "Entangled Lives," 530–40; Rebecca Gowland and Sophie L. Newman, "Children of the Revolution: Childhood Health Inequalities and the Life Course During Industrialisation of the 18th to 19th Centuries," in *Children and Childhood in the Past* (Gainesville: University Press of Florida, 2018).

33. Hertzman, "Biological Embedding in Historical Perspective," 17160–167.

34. Thayer and Kuzawa, "Biological Memories of Past Environments," 798–803.

35. See, for example, Margaret Lock, "The Epigenome and Nature/Nurture Reunification: A Challenge for Anthropology," *Medical Anthropology* 32, no. 4 (2013): 291–308.

36. Meloni, "Social Brain," 5.

References

Agarwal, Sabrina C., and Bonnie A. Glencross. *Social Bioarchaeology*. Chichester, UK: Wiley-Blackwell, 2011.

Budgeon, Shelley. "Identity as an Embodied Event." *Body and Society* 9, no. 1 (2003): 35–55.

Butler, Judith. *Bodies That Matter: On the Discursive Limits of Sex*. London: Routledge, 1993.

Cole, Simon A. "A Surfeit of Science: The 'CSI Effect' and the Media Appropriation of the Public Understanding of Science." *Public Understanding of Science* 24, no. 2 (2015): 130–46.

Cole, Simon A., and Rachel Dioso-Villa. "Investigating the CSI Effect: Media and Litigation Crisis in Criminal Law." *Stanford Law Review* 61 (2010): 1335.

Crossland, Zoe. "Evidential Regimes of Forensic Archaeology." *Annual Review of Anthropology* 42 (2013): 121–37.

Dror, Itiel E. "Biases in Forensic Experts." *Science* 360, no. 6386 (20 April 2018): 243.

Fausto-Sterling, Anne. "The Bare Bones of Sex: Part 1—Sex and Gender." *Signs: Journal of Women in Culture and Society* 30, no. 2 (2005): 1491–527.

Ferrándiz, Francisco. "Exhuming the Defeated: Civil War Mass Graves in 21st-Century Spain." *American Ethnologist* 40, no. 1 (2013): 38–54.

Fowler, Gillian, and Tim Thompson. "A Mere Technical Exercise? Challenges and Technological Solutions to the Identification of Individuals in Mass Grave Scenarios in the Modern Context." In *Human Remains and Identification: Mass Violence, Genocide and the "Forensic Turn,"* edited by Élisabeth Anstett and Jean-Marc Dreyfus, 117–41. Manchester, UK: Manchester University Press, 2015.

Geller, Pamela L. *The Bioarchaeology of Socio-Sexual Lives: Queering Common Sense About Sex, Gender, and Sexuality*. Zug, CH: Springer International, 2016.

———. "Conceiving Sex: Fomenting a Feminist Bioarchaeology." *Journal of Social Archaeology* 8, no. 1 (2008): 113–38.

Gowland, Rebecca. "Entangled Lives: Implications of the Developmental Origins of Health and Disease Hypothesis for Bioarchaeology and the Life Course." *American Journal of Physical Anthropology* 158, no. 4 (2015): 530–40.

Gowland, Rebecca, and Christopher Knüsel. *Social Archaeology of Funerary Remains*. Oxford, UK: Oxbow Books, 2006.

Gowland, Rebecca, and Sophie L. Newman. "Children of the Revolution: Childhood Health Inequalities and the Life Course During Industrialisation of the 18th to 19th Centuries." In *Children and Childhood in the Past*. Gainesville: University Press of Florida, 2018.

Gowland, Rebecca, and Tim Thompson. *Human Identity and Identification*. Cambridge: Cambridge University Press, 2013.

Hertzman, Clyde. "Putting the Concept of Biological Embedding in Historical Perspective." Supplement, *Proceedings of the National Academy of Sciences* 109, no. S2 (16 October 2012): 17160–167.

Hollimon, Sandra E. "Sex and Gender in Bioarchaeological Research: Theory, Method, and Interpretation." *Social Bioarchaeology*, 2011, 147–82.

Joyce, Rosemary A. "Archaeology of the Body." *Annual Review of Anthropology* 34 (2005): 139–58.

Karkazis, Katrina, Rebecca Jordan-Young, Georgiann Davis, and Silvia Camporesi. "Out of Bounds? A Critique of the New Policies on Hyperandrogenism in Elite Female Athletes." *American Journal of Bioethics* 12, no. 7 (2012): 3–16.

Kuzawa, Christopher W. "Developmental Origins of Life History: Growth, Productivity, and Reproduction." *American Journal of Human Biology* 19, no. 5 (2007): 654–61.

Landecker, Hannah. "Food as Exposure: Nutritional Epigenetics and the New Metabolism." *BioSocieties* 6, no. 2 (2011): 167–94.

Landecker, Hannah, and Aaron Panofsky. "From Social Structure to Gene Regulation, and Back: A Critical Introduction to Environmental Epigenetics for Sociology." *Annual Review of Sociology* 39 (2013): 333–57.

Laqueur, Thomas. *Making Sex: Body and Gender from the Greeks to Freud*. Cambridge, MA: Harvard University Press, 1990.

Lock, Margaret. "The Epigenome and Nature/Nurture Reunification: A Challenge for Anthropology." *Medical Anthropology* 32, no. 4 (2013): 291–308.

Low, Felicia M., Peter D. Gluckman, and Mark A. Hanson. "Developmental Plasticity, Epigenetics and Human Health." *Evolutionary Biology* 39, no. 4 (2012): 650–65.

Maples, William R. "The Practical Application of Age-Estimation Techniques." In *Age Markers in the Human Skeleton*, edited by Mehmet Yaşar İşcan, 319–24. Springfield, IL: Charles C. Thomas, 1989.

Meloni, Maurizio. "The Social Brain Meets the Reactive Genome: Neuroscience, Epigenetics and the New Social Biology." *Frontiers in Human Neuroscience* 8 (2014): 309.

Meskell, Lynn, and Robert W. Preucel, eds. *Companion to Social Archaeology*. Malden, MA: Blackwell, 2004.

Nakhaeizadeh, Sherry, Itiel E. Dror, and Ruth M. Morgan. "Cognitive Bias in Forensic Anthropology: Visual Assessment of Skeletal Remains Is Susceptible to Confirmation Bias." *Science and Justice* 54, no. 3 (2014): 208–14.

Nakhaeizadeh, Sherry, Ian Hanson, and Nathalie Dozzi. "The Power of Contextual Effects in Forensic Anthropology: A Study of Biasability in the Visual Interpretations of Trauma Analysis on Skeletal Remains." *Journal of Forensic Sciences* 59, no. 5 (2014): 1177–83.

Nakhaeizadeh, Sherry, Ruth M. Morgan, Carolyn Rando, and Itiel E. Dror. "Cascading Bias of Initial Exposure to Information at the Crime Scene to the Subsequent Evaluation of Skeletal Remains." *Journal of Forensic Sciences* 63, no. 2 (2018): 403–11.

Oudshoorn, Nelly. *Beyond the Natural Body: An Archaeology of Sex Hormones*. London: Routledge, 1994.

Shilling, Chris. *The Body and Social Theory*. London: Sage, 2012.

Sofaer, Joanna R. *The Body as Material Culture: A Theoretical Osteoarchaeology.* Vol. 4. Cambridge: Cambridge University Press, 2006.

Sofaer Derevenski, Joanna R. "Sex Differences in Activity-Related Osseous Change in the Spine and the Gendered Division of Labor at Ensay and Wharram Percy, UK." *American Journal of Physical Anthropology* 111, no. 3 (2000): 333–54.

Sørensen, Marie Louise Stig. *Gender Archaeology.* Hoboken, NJ: Wiley, 2000.

Thayer, Zaneta M., and Christopher W. Kuzawa. "Biological Memories of Past Environments: Epigenetic Pathways to Health Disparities." *Epigenetics* 6, no. 7 (2011): 798–803.

Thompson, Tim, Daniel Jiménez Gaytan, Shakira Bedoya Sánchez, and Ariana Ninel Pleitez Quiñónez. "Forensic Anthropology: Whose Rules Are We Playing By?—Contextualizing the Role of Forensic Protocols in Human Rights Investigations." In *War Crimes Trials and Investigations*, 59–80. London: Palgrave Macmillan, 2018.

Tuller, Hugh. "Identification versus Prosecution: Is It That Simple, and Where Should the Archaeologist Stand?" In *Disturbing Bodies: Perspectives in Forensic Anthropology*, edited by Zoe Crossland and Rosemary A. Joyce, 85–102. Santa Fe, NM: School for Advanced Research Press, 2015.

Turner, Bryan S. *The Body and Society: Explorations in Social Theory.* Oxford, UK: Blackwell, 1984.

Walker, Phillip L. "Sexing Skulls Using Discriminant Function Analysis of Visually Assessed Traits." *American Journal of Physical Anthropology* 136, no. 1 (2008): 39–50.

6.
Becoming More Biological
Ruth Ozeki and the Postgenomic Ethnoracial Novel

Patricia E. Chu

The last great battle over racism will be fought not over access to a lunch counter, or a hotel room, or to the right to vote, or even the right to occupy the White House; it will be fought in a laboratory, in a test tube, under a microscope, in our genome, on the battleground of our DNA. It is here where we, as a society, will rank and interpret our genetic difference.
—Henry Louis Gates Jr., "The Science of Racism"

In Ruth Ozeki's *My Year of Meats* (1998), a young filmmaker gets a contract to work for the company BEEF-X on a Japanese marketing campaign for a TV show called *My American Wife!*, which will be broadcast in Japan for the purposes of getting Japanese housewives to buy more American beef. For each episode, Jane Takagi-Little chooses a family from a different small town in America. Jane and her crew film the family, their town, and the wife/mother making her family's favorite recipe. Housewives in Japan can follow along, copy the list of ingredients, and replicate the dish while learning about American culture. Jane's mother is from Japan, and Jane is fluent in Japanese; she is able to work with a Japanese film crew and with the home office in Japan running the campaign for their American partners while talking to people in small-town America.

In this essay I read *My Year of Meats* as a postgenomic text[1] through which we might understand shifts in the American ethnoracial novel in terms of the biocultures that have arisen post–Human Genome Project. Lennard J. Davis and David B. Morris write in their "Biocultures Manifesto" that "at the beginning of the twenty-first century, we make a new (but perhaps in a while old) and counterintuitive (but perhaps destined to be commonplace) proposal: that culture and history must be rethought with an understanding of their inextricable, if highly variable, relation to biology."[2] Although they don't specifically address the rise of the mapping of the human genome and the biotechnology industry that has arisen around the possibilities of genomic research, their call has obvious relevance for the contemporary moment, with its increased emphasis on the significance of and possibilities for knowledge about one's biological identity—that is, one's individual genome and its probabilities and the relation one might have as an individual to a population based on molecular biological identity. As Gates's words illustrate, there is a sense in which the era of the genome may make the traditional battlegrounds of the Civil Rights Movement obsolete. This bioculture is a fraught environment for authors of the American ethnoracial novel, because the historical mission of such authors has been to challenge biological determinism in identity.

My Year of Meats explores Jane's identity against a backdrop that we can associate with the entrance of molecular biotechnology into economic, cultural, and political life: "big" biotechnological agriculture and pharmaceuticals. Jane learns about mass feedlots and slaughterhouses and that she is a DES daughter, a woman exposed in the womb to diethylstilbestrol, a synthetic hormonal drug meant to prevent miscarriages. Because of this, she is at higher risk of some cancers and infertility disorders. In setting a traditional ethnoracial identity amid other kinds of biologized identities, Ozeki draws attention to how the categories and inequalities traditionally addressed in the ethnoracial novel cannot fully capture the new biocultural context.

Shifts in US government policy involving scientific research, global trade, and financial investment enabled biotechnology to begin its rise to significance as a driver of economic profit and generator of surplus value. As Melinda S. Cooper details in *Life as Surplus*, it is not merely coincidence that Ronald Reagan presided over the early implementation of both a neoliberalist economic policy and a life sciences program, for the two share "a common ambition to overcome the ecological and economic limits to growth associated with the end of industrial production."[3] The industries (plastics, fabrics, agriculture, pharmaceutical, and petrochemical) that invested in genetic technologies were those that were running up against the high costs of the environmental waste caused by industrial production. Side effects of Green Revolution exports, the toxicity of new drugs such as thalidomide, and public outcry about chemical trials of pharmaceuticals and about increasingly visible ecological pollution had led to increased regulation of production, which raised costs and lowered profits. Industrial production also depended on oil, a nonrenewable resource. These companies began to "relocate beyond the limits of industrial production—in the new spaces opened by molecular biology..., market[ing] themselves as purveyors of new, clean, life science technologies. By the early 1980s all of the major chemical and pharmaceutical companies had invested in the new genetic technologies, either through licensing agreements with biotech startups or by developing their own in-house research units."[4] Government policies continued to shift development funds away from basic public research to private, for-profit commercial projects. By the Clinton era of the late 1990s, with its upturn in the investment market (the basis of the new kinds of financialization that became the foundation of biotech corporations) and its more full-blown neoliberal economic policies, "life becomes, literally, annexed within capitalist processes of accumulation."[5] Here, "life" refers not simply to life-forms but also to life processes and life materials. As Cooper says,

> What counts here is the variable source code from which innumerable life forms can be generated, rather than the life form per se. Hence the biological patent allows one to own the organism's *principle of generation* without having to own the actual organism. In the age of postmechanical reproduction the point is . . . to generate and capture production itself, in all its emergent possibilities. Its success is dependent on the constant transformation of (re)production, the rapid emergence and obsolescence of new life forms, and the novel recombination of DNA rather than the mass monoculture of standardized germplasm.[6]

Cooper asks, "What finally becomes of the critique of political economy in an era in which biological, economic, and ecological futures are so intimately entwined?"[7] I argue that this is a question we might ask about the artist and writer of the ethnoracial text in the contemporary moment and that Ozeki's *My Year of Meats* is one of the early attempts to consider this context—this bioculture—in the frame of the ethnoracial. What emerges is an ethnoracial novel that does not look like what we have been expecting but one that responds to a new understanding of biological identity.

Stevens and Richardson define the postgenomic age temporally as the period after the completion of human genome sequencing and technologically as the introduction of whole genome methods such as databases and biobanks, sequencing technologies, and big data in genome-wide association studies.[8] As the concepts and promise of molecular biology and biotechnology permeate social, political, and economic life, our understandings of self become more and more biological. In an early cultural consideration of the Human Genome Project, Ian Hacking categorizes Philip Roth's *The Human Stain* as a "pre-genetic" novel because of its ideology of choice and performative ethnic identity. Post–Human Genome Project, he describes the rise of a different sense of identity: "We all carry an enormous mix of inheritance, and the greater the extent to which a person's recent forbears came from geographically disparate parts of the globe, the greater the possibilities for picking out and identifying with this or that distinct strand."[9] He imagines a future political scenario in which a wealthy Brazilian capitalist who looks White wishes to run for office and needs to declare himself, despite his wealth, a man of the people. He has himself genetically tested, and the results come back that he is more African than Portuguese. His political party publishes them. Hacking writes, "The opposition repeats to no effect that this hardly distinguishes him from anyone else in Brazil, that he is nothing but a playboy from Sao Paulo whose grandparents were smart enough to become very rich."[10]

Hacking was prescient here. Ancestry.com runs an advertisement wherein a man named Kyle Merker talks about how "we grew up thinking we were German," and so he spent his childhood dancing German dances and wearing lederhosen. After learning that he was really more Irish than German through his Ancestry.com test, he has given up his German practices and is now "happy" in an Irish kilt.

This begs a host of questions even beyond the fact that companies such as Ancestry.com are not measuring actual "race genes" or genes that are always associated with particular races, since such genes do not exist. Instead, the tests look for "single nucleotide polymorphisms" (or SNPs, which have nothing to do with phenotypical race) that arise by random mutation and are passed on to offspring, meaning that they tend to be most common at the point of origin of the population in which they occurred and spread outward slowly, thus becoming more diffused. The idea behind the swab tests, then, is to draw an association between particular clusters of SNPs and geographic location. But the nature of SNPs is such that, as John Dupré explains, while the tests do predict fairly well that a person who identifies as African American will have SNPs that originated in West Africa, those same SNPs will be found in people who do not so identify because SNPs diffuse outward as a result of interbreeding.[11] Genetic ancestry testing of maternal and paternal lineages traces one line of descent out of the many lines each individual has. Going back fourteen generations, each person can have as many as 16,384 direct ancestors. In other words, test results do not represent genetic makeup—"an individual might have more than 85% western European 'genomic' ancestry but still have a West African mtDNA or NRY heritage."[12] Finally, as Deborah A. Bolnick explains, in designing and interpreting the tests we have decided to make some geographical divisions and points in ancestral history more relevant than others. The distant ancestors of all humans lived in Africa.[13]

The national boundaries and cultural identities the Ancestry.com ad uses to differentiate between Germans and Irish as though they are absolute are relatively recent. In the context of recent debates on European nationalism, historian Patrick J. Geary criticizes the idea that European peoples have been and remained stable groups of clearly delineated social classifications.[14] He points out that the ancestral history of Germanic and Celtic peoples is actually quite mingled:

> The Celts, another Indo-European people . . . spread[] from what is today Czechoslovakia, Austria, and southern Germany and Switzerland to Ireland in the sixth century b.c.e., pushing back, absorbing, or eradicating the indigenous European population until the only survivors were the Basques of southern France and northern Spain. From the first century b.c.e., Germanic peoples began pushing the Celts from the east to the Rhine, but they and the Celts confronted a different invader: the expanding Roman Empire, which conquered and Romanized much of Europe as it did Asia Minor and North Africa.[15]

In other words, the difference between Irish and German heritage offered by Ancestry.com and taken up so seriously by Kyle Merker does not refer to peoples or cultures in any historically or biologically accurate way. Irish and German geographies were not "Irish" or "German" as we think of them today.

In any case, what does it mean that Kyle Merker or someone like him would, after many years, stop participating in family practices or quit activities with ethnic associations because of a DNA test? As with Hacking's imaginary politician, his life history and experiences are swept aside. And even Hacking's idea that greater knowledge of biological identity might offer "greater . . . possibilities for picking out and identifying with this or that distinct strand" doesn't seem to lead to an expansion of possibilities. If he is X, he can't be Y. Why is Merker "happy" in his kilt rather than devastated to lose his previous social network? Why is the conclusion of the ad not a mixture of Irish and

German clothing and practices? If he is happy in his kilt now, presumably because his knowledge matches some deep biological truth of his body, what does it mean that he was not previously unhappy in his lederhosen? Ancestry.com seems to find explanation unnecessary.

The Ancestry.com ad echoes what Gerlach and his colleagues call the rise of a new kind of subjectivity: biosubjectivity, wherein we believe that "truth exists not on the material plane of the body but within a microcosmic regime of genetic information."[16] This information "has the power to locate a person's position within different social fields, to alter one's perception of oneself, and to shape the pathways of their regulations and mobility."[17] Would Merker's dance-group companions reject him if they were shown a copy of his test results, and if so (or if not), why? Why does this ad seem to reinforce the significance and immutability of ancestry rather than its insignificance or uncertainty if the narrative of the ad doesn't bother to hide the fact that an "Irish" person could be so good at being "German" for so many years?

How is the shift in biosubjective ethnoracial identifications and practices we see in the proliferation of a range of genomic and biotechnological practices such as recreational ancestry searches affecting a literary form whose politics has traditionally *challenged* the immutability and limits of biological identities and classifications? Indeed, can there still be American ethnoracial literature, in the sense of a literature that explores the constructions, experiences, and politics of identities, if the answer is simply a printout and a mandatory switch of clothing?

In contrast, in an earlier moment in American literary history, Frances Watkins Harper's 1892 eponymous heroine Iola Leroy lives the privileged life of a Southern slaveholder's daughter until she finds out she is ancestrally and legally Black. Despite her slaveholding heritage, she is enslaved and then freed. Afterward, she chooses not to live as White even though she looks White and was raised as a White woman. Instead, she chooses to live as Black and work for Black liberation during and after the Civil War because she understands the categorical politics of race as more than simply ancestry. Iola is offered marriage by a White man who says they can move away and never tell anyone she has Black ancestry. He makes a cultural argument that having been raised and educated as a White woman she can never really fit into a Black community and that her social place is rather (conveniently for him) to be the wife of a man like himself. She responds, "It was through their unrequited toil that I was educated, while they were compelled to live in ignorance. I am indebted to them for the power I have to serve them. I wish other Southern women felt as I do."[18] Here, the knowledge of ancestry provokes a very complicated consideration of identity—Iola marks herself as a "Southern" woman and acknowledges the part of her identity that is "White" as neither merely biological nor merely cultural but also political and economic. Other Southern women could feel the way she does despite not being Black in the way she has learned she is, and that kind of Black identification could affect their White identity, because it would be an identification that emerged from their understanding of where their Whiteness—which here is not merely ancestral or biological—came from.[19] The plot and developing themes of the novel derive energy from a constant consideration of what Iola's new knowledge of her race could mean. Whiteness and Blackness both factor into her decisions about how to identify, and identification is political as well as individualist-subjective. Unlike the contemporary case of Kyle Merker, Iola's ancestral knowledge expands the possibilities of identification rather than switching one on and another off, and this expansion both drives and is driven by the discursive capacities of literature and language as Harper deploys them. At the same time, and this is the core mission of the traditional ethnoracial novel, Iola's capacities (despite her "biology") are shown to be as unlimited and undetermined as any White person's because she was raised as a White person, with all the resources and expectations of a White family. During the period of Reconstruction, Harper makes a call for the irrelevance of the biological in personal life and civil democracy.

About a century later, midway through the Human Genome Project's mapping of a human, Ozeki too created a mixed-race character. Hazel Carby has argued that the mulatto figure of nineteenth-century American literature, and specifically in *Iola Leroy*, is a "narrative figure" having "two primary functions: as a vehicle for an exploration of the relationship between the races and, at the same time, an expression of the relationship between the races . . . , a narrative device of mediation."[20] The "American half" of a protagonist warring in agony with the "Japanese half," the most famous example of which is the fragmenting bodies and minds in John Okada's *No-No Boy* (regardless of actual biological parentage—Ichiro and the other no-no boy characters have Japanese mothers and fathers), is a trope in Asian American literature. Frank Chin denounces a study in which students were asked to list their American qualities in one column and their Asian qualities in the other. Maxine Hong Kingston picks this up in *Tripmaster Monkey* as well. There, the criticism is how such a mode of thought makes the idea of the Asian American impossible because of the stereotype of the Asian as inassimilable to America and to democratic practices and also enforces the idea that the two cultures are entirely different. Meanwhile, White ethnics get to "hyphenate" quite comfortably. This can allow culture to cover for state racism. For instance, a character in *No-No Boy* realizes that members of his parents' generation may have felt forced to cling to their previous identity because immigration laws specifically made them ineligible for citizenship. More recent works such as Han Ong's *Fixer Chao* mock the trope, pointing to how it has become an internal and external stereotype that covers up both state and economic racism. In *My Year of Meats,* things are less figurative.

Indeed, far from creating a heroine who belies biology or a text that metaphorizes biology in its account of cultural identity, Ozeki has Jane embrace a biological identity, even specifically crafting it (with a twist) from an antique racist primer: *Frye's Grammar School Geography* by Alexis Everett Frye (a real book not invented by Ozeki).[21] Jane becomes fascinated with the book as a child when she finds it in her school library. In this primer, young Jane reads, "If we were to travel all through the countries, we should see many different classes of people. We may divide them into five great groups called races...." She learns that "all of Africa south of the Sahara is the home of the *black* or *Negro race*.... Such natives are very ignorant ...," and "red men resemble the black savages of middle Africa.... They wear but little clothing and use about the same kinds of weapons. They hunt and fish and lead a lazy, shiftless life.... The people of the yellow race living on the island of Japan have made more progress than any other branch of the race.... They have been wise enough to adopt many of the customs of the white race."[22]

Disregarding the racism ("even as a kid I knew there was something very wrong with this picture of the world—after all I had gone to the library... searching for amalgamation, not divisiveness"[23]), Jane sees her mixed-race status as the beginning of a "breeding project" that could amalgamate the separate categories in a baby that is an "embodied United Nations."[24] "I learned what I needed: a mate who was black, brown, or red to go with my white and yellow. At the very least, I was aiming for three out of five."[25] Challenging the geographical organization of *Frye's Geography*, she argues against nativism and imagines herself as a commercial experiment ("prototype").

> All over the world, native species are migrating, if not disappearing, and in the next millennium the idea of an indigenous person or plant or culture will just seem quaint. Being half, I am evidence that race, too, will become relic. . . . I feel brand-new—like a prototype.²⁶

When her now ex-husband Emil, an African engineer from Zaire whom she meets in Kyoto, learns about what she calls, looking back, "my experiment in biotech," he jokes that "[he] will gladly be the genetic engineer of [their] love, but perhaps [they] ought to marry first."²⁷ Here we see Ozeki tracking a genealogy of racial thinking across a century, highlighting how race as conceived in something like *Frye's Geography* discursively converts to a new biology at the beginning of the twenty-first century. Jane adopts her biological engineering project with a Kyle Merker–like enthusiasm and faith in the categories that she had already rejected as a child. Here the antidote for the biologically based racism that shaped her childhood is not an antibiological narrative but the adoption of a biological subject position and a biotechnological solution. Jane cannot get pregnant with Emil. Frustrated, she comments with bitter, ironic humor that her adoption of "mulatto" as a biological self-description that incidentally describes her personality seems to be extending into its older racist meaning:

> I have thought of myself as mulatto (half horse, half donkey—i.e., a "young mule"), but my mulishness went further than just stubbornness or racial metaphor. Like many hybrids, it seemed, I was destined to be nonproductive.²⁸

After testing she discovers that she has a precancerous neoplasia in her cervix and a deformed fallopian tube. But this, too, as I will discuss later, is a bioracial identity insofar as it is quite possible that an Asian woman in Quam, Minnesota (Jane's mother), looked like she needed to be treated with DES (the likely cause of the deformation).

Biological identity continues to expand throughout the book in ways that are antithetical to some of the traditions of the ethnoracial novel. I should make it clear that I am not arguing here that Ozeki is writing a version of the Ancestry.com ad. Rather, I see the text as maneuvering within the new discourses of biology and race to understand the political implications of identity in the genomic age for the genre of the ethnoracial novel.

Ozeki signals her sense of the inadequacy of past tropes by undercutting what has been a de rigueur moment in Asian American identity stories: the "Where are you from?" moment.²⁹

At the pancake breakfast where we had been filming, a red-faced veteran from WWII drew a bead on me and my crew, standing in line by the warming trays, our plates stacked high with flapjacks and American bacon.

"Where are you from, anyway?" he asked, squinting his bitter blue eyes at me.

"New York," I answered.

He shook his head and glared and wiggled a crooked finger inches from my face.

"No, I mean where were you *born*?"

"Quam, Minnesota," I said.

"No, no. . . . *What* are you?" He whined with frustration.

And in a voice that was low but shivering with demented pride, I told him, "*I . . . am . . . a . . . fucking . . . AMERICAN!*"[30]

Here anyone familiar with Asian American novels or Asian American politics or living as Asian American recognizes the question and the assumptions underlying it. But Ozeki makes Jane the one in this interaction who is a "demented" nationalist. This encounter takes place shortly after Jane and her crew land at the local airport in the midst of Gulf War fever as soldiers are leaving on deployment. Her outrage is thus tainted by what this claim to being "a . . . fucking . . . AMERICAN" means. In earlier Asian American novels, taking a stand against the stereotype of the perpetual foreigner and against the claim that Asian American history is shorter than that of some White immigrant/hyphenated groups was prioritized despite the problematic implications of claiming one's identity as American and thus accepting imperial or domestic racist politics in making that claim.[31] But here, Ozeki questions Jane's momentary willingness to be as militantly American as someone with war-hungry, nationalist Gulf War fever. Ozeki then drops the issue of "claiming Americanness." Jane encounters the assumption that she is Japanese only when she is with her (actually Japanese) film crew and before she speaks in English. This has the effect of making the trope seem outmoded.

There are other ways in which Ozeki moves away from the traditional tropes of ethnoracial identity and protest to adopt those that are more biological and less overtly those of naming or categorization. First, it is notable that *My Year of Meats* is unusual in being a Japanese American novel with no internment narrative. In many earlier Japanese American texts, the loss of partly personal, partly public history is indexed by the internment, something unacknowledged in much of American history and often such a painful subject in Japanese American families that children's questions go unanswered. But this is a Japanese American novel that does not mention internment. Jane's mother is a war bride, not an internee. Jane's biological connection to her racial ancestry is primarily a problem of personal biological production, not of reproduction of culture or recovery of history. She is not an internee's daughter but a DES daughter with a damaged fallopian tube and a high risk of early-onset, rapid-progression cancers. Jane wonders whether her mother might have seemed a natural candidate for DES because as a small Asian woman in Minnesota she looked like someone who would have a difficult pregnancy; she eventually learns that her mother had four miscarriages, one of the many conditions that the pharmaceutical company aggressively marketed its drug as preventing (this claim was later shown to be untrue). Such marketing resulted in DES being distributed to millions of women in America. In other words, significant biology here is not traditionally racial.

Jane is the daughter, on the other side of her family, of a father who was a US army botanist sent to Hiroshima to research the effects of nuclear radiation. "They were kind of checking up on their handiwork—you know, looking at people and monstrose plant mutations—to see if we should drop an A-bomb on Korea. Died of cancer and I've always wondered whether there's some connection."[32] This is how Jane begins the history of her family. When her boyfriend Sloan asks how her parents met, Jane continues, "'The old story. Ma was a prostitute on the streets of Tokyo, Dad was a GI—' 'No shit. . . . Really?' 'Nope.'" Here, "the old story" of American wars in Asia (what in a traditional ethnoracial text might be used to thematize American imperialism using a mulatto figure in the way that Carby describes) is explicitly invoked in order to reject it as a foundational mode of identity for Jane. Both her parents are carriers of biological meaning that is not traditionally racial. And although the novel follows the genre's demands for the reclamation or recontextualization of history, Jane's history is biocultural, not strictly ethnoracial; she learns she is a DES daughter, and in her travels and research for the meat industry she also learns about big agriculture's use of DES and other hormones. This history then circles back to another part of her family history on the Little side.

> DES changed the face of meat in America. Using DES and other drugs, like antibiotics, farmers could process animals on an assembly line, like cars or computer

chips. . . . It was happening everywhere, the wave of the future, the marriage of science and big business. . . . My grandparents, the Littles, lost the family dairy farm to hormonally enhanced cows, and it broke their hearts and eventually killed them. But I'd never understood this before.[33]

Jane's family history encompasses a larger history. The "marriage of science and big business" is brokered by the neoliberal imperative for endless growth. This imperative results in a political economy in which, as various critics describe it, "the market should be the organizing principle for all political, social, and economic decisions"; "the economy should dictate its rules to society, not the other way around"; and "the general idea that society works best when the people and the institutions within it work or are shaped to work according to market principles,"[34] creates a system in which economic efficiency is unquestioned and there are no limits set on the market or on what can be financialized.[35]

It is important to emphasize Cooper's history of intertwined science and history here. This is an economy interarticulated with the establishment of biotechnology. DES, hormones, and antibiotics as part of the business of selling meat in *My Year of Meats* index this. Jane and her grandparents are related through DES as significantly as they are through a family tree. The elder Littles are economically surplused in a bioculture that efficiently manipulates the workings of biological life when their cows are not as efficient because the Littles will not give them hormones. If we understand the relation of Jane and her grandparents as also biopolitical according to Cooper's analysis rather than strictly familial-biological according to a model of racialized descent, we can see how Ozeki is reshaping the structure of biological identity within ethnoracial identity. On the one hand, descent is key to the plot in terms of Jane's discovery of family history. On the other hand, a more "sideways" biological relation—through DES—is also established. This other kind of relation, characterized by relation *through biology* rather than *biologically*, parallels the biotechnological breakthrough of recombinant DNA, which "mobilizes the transversal processes of bacterial combination rather than the vertical transmission of genetic information."[36] The older model of descent, so central to the ethnoracial novel outlining a group identity (indeed, one might argue, a model central to the novel genre itself), is still present but alongside a new consciousness of the self as biological in another way. Rachel C. Lee writes that Ozeki uses a model of "entangled populations" that pulls the reader "beyond the narrowly defined species-being to which Asian Americans have thought themselves to belong."³⁷

To better understand the changed quality of Ozeki's "species-being," the ethnoracial group, let us consider one of the main ways the novel has been read: as an ecofeminist exploration of how Americans produce and consume meat. These readings cast *My Year of Meats* as drawing parallels between the treatment of women and of animals in a patriarchal society to examine the reduction of women to the merely biological functions they perform and to highlight the violence of Americans that arises from their capacity to internalize the rationality of meat production and consumption. Ueno's job as marketer of meat slides into his physical violence against his wife Akiko, whose "meat duties" are to watch, cook, and serve the weekly recipe from *My American Wife!* but also to force herself to eat meat so that she will gain weight, have a normal menstrual cycle, and be able to bear Ueno's child. DES's history of being used on both cows and women to control reproduction and produce more milk and meat is used to emphasize this point.³⁸

These are illuminating readings, but I'd like to tease out some additional aspects of the novel. The racist *Frye's Geography*, as we have seen, reads differently to Jane than it might in a traditional ethnoracial novel. As the text describes itself:

> In this book, man is the central thought. Every line of type, every picture, every map, has been prepared with a single purpose, namely, to present the *earth as the home of man*,—to describe and locate the natural features, climates and products that largely determine his industries and commerce, as well as his civic and other relations,—thus bringing REASON to bear on the work.³⁹

Jane specifically notes that she doesn't care about Frye using "man," calling that objection an "intraspecific quibble." Instead, she says, "The conflict that interests me isn't *man* versus *woman*, it's *man* versus *life*. Man's REASON, his industries and commerce, versus the entire natural world."⁴⁰ This passage has been used in support of an ecofeminist association of women with the nature that is acted on by science/men. In this reading, Ozeki's clear rejection of DES would extend to all biotechnology and establish the kind of strict divide about which Donna Haraway has taught us to be skeptical.⁴¹ But note that it is the unredeemable Ueno who insists on a "natural" model of eugenic patriarchal biological descent ("I want *my own children*. Mine. Do you hear? *Mine*. Not some bastard of a Korean whore and an idiot American soldier. I want *my* genes in *my* child. That's the point. *Mine!*"⁴²) that is completely contrary to Jane's play with biology and her embrace of mixtures such as a woman speaking masculine Japanese, androgynous dress, adoption, immigration, racial mixing, and being one's own biological and political prototype: in short, Jane, not the patriarchal Ueno, appreciates unnatural (because not based on genetic descent models?), "sideways," and recombinant biology and culture.

This is not to say that Ozeki writes a celebration of all biotechnology. Throughout the novel, Jane encounters other instances of the way the biotechnological political economy impacts other bodies: the Purcells tell her about how eating chicken parts raised Mr. Purcell's voice and gave him breasts; her director Oda is hospitalized with anaphylactic shock from antibodies in schnitzel cooked by one of the American wives; and when the team visits a cattle slaughterhouse and ranch, they meet Rosie, a five-year-old girl with the breasts of a full-grown woman and early menstruation as a result of hormone use in the cattle feed. Ozeki draws the line at the marriage of biology and business. Gale's use of hormones and antibiotics, some illegal, is "necessary" if he is to keep the ranch competitive, but Rosie pays the price. Another way to see how Ozeki draws this line is to compare the way Gale and Grace, one of the American wives, use the phrase "The math just don't work out."⁴³ This can be a neoliberal phrase: if the math works out, then that is all that matters and any social good must also be advocated for in terms of its economic value (i.e., "healthcare for all actually saves money"). But Ozeki makes a differentiation. Whereas Gale uses the phrase to justify using drugs on the cows ("profit's so small these days you gotta deal in volume...; without the drugs we'd be finished"),⁴⁴ Grace uses it to outline a larger problem with valuing lives. She describes reproduction as the basis of the problems "we talk about": "environment, the economy, [and] human rights,"⁴⁵ although no one will discuss population. She and her husband adopt most of the large family they always wanted. Here, we might see what Zygmunt Bauman, Angela Y. Davis, and Henry Giroux describe as a defining characteristic of neoliberalism: the idea that some populations are surplus and disposable in the wake of the state's inability and unwillingness to maintain a social welfare safety net for those who cannot contribute to economic growth by producing or consuming properly.⁴⁶ As Grace says in her interview with Jane, "Havin' babies is going to be the big topic of the millennium—who gets to do it, who's still even capable of doing it."⁴⁷ Ominously, what business does is clearly illustrate where those rights come from: profit. Cattle (not cows in this view, Jane points out) for instance—are fecund (versus fertile) via intersection with the market but are understood as life *matter* rather than life-*form*.

Biologically recombinant sideways generation and affiliation are also aesthetically part of shifting the range of the ethnoracial in *My Year of Meats* while challenging identity structures like those of Ancestry.com—biologically based structures that are, after all, both vertical and rigid. Take the two intertexts of *My Year of Meats: Frye's Geography* and the Japanese classic text *The Pillow Book of Sei Shonagon*. One of these is a text of genetic descent—a Japanese American looks back to a Japanese text in a way that has become quite an expected pattern in ethnoracial literature. The other is a text whose incorporation only invokes biotechnological recombination, since *Frye's Geography* has no direct relation to Jane's racial group identity and must be read against the grain to support the model of identity she adopts. Jane's primary medium—film—parallels postgenomic sequencing work in such a way that, for both, editing is central to creation and is a method for ignoring the set genre—the supposedly natural—of previous generations. "Editing is what counts," she says.⁴⁸ Her interpretation of the central conflict of *Frye's Geography* as *"man's* REASON, his industries and commerce, versus the entire natural world" might be read, then, not as a nostalgic desire for natural divisions and groupings but in the way we have seen Jane read *Frye's Geography* throughout the novel: for its recombinant possibilities set against rigidly naturalized ethnoracial identity. In short, the new ethnic novel pits biology against biology, genetics against genomics and molecular biology.

The Ancestry.com ad featuring Kyle Merker might lead us to worry about the state of the ethnoracial story in the age of the genome. But Ozeki's novel illustrates what Richard Doyle has said about the way the life sciences, rather than reducing the possibilities for narrative to a printout, leaves us with "nothing but story, nothing but information. . . . A researcher will begin with the end of narrativity, with the idea that 'all there is' to know about an organism can be found in its electronic database sequence, the notion that such information is the timeless and perhaps indestructible essence of the organism. But she will not stop there: the narrative of life will now be an exegetical one where theorists scan databases and . . . produce new knowledge [and] new stories of organisms"⁴⁹ and of biologized identity.

Notes

1. As I will discuss later, this follows Hacking's discussion of a Roth novel as "pre-genetic." Ian Hacking, "Genetics, Biosocial Groups, and the Future of Identity," *Daedalus* (Fall 2006): 92.

2. Lennard J. Davis and David B. Morris, "Biocultures Manifesto," *New Literary History* 38 (2007): 411–18.

3. Melinda Cooper, *Life as Surplus: Biotechnology and Capitalism in the Neoliberal Era* (Seattle: University of Washington Press, 2008), 11.

4. Ibid., 21–22.

5. Ibid., 19.

6. Ibid., 24.

7. Ibid., 20.

8. Hallam Stevens and Sarah S. Richardson, "Beyond the Genome," in *Postgenomics: Perspectives on Biology After the Genome* (Durham: Duke University Press, 2015), 3.

9. Hacking, "Future of Identity," 92.

10. Ibid., 93.

11. John Dupré, "What Genes Are and Why There Are No Genes for Race," in *Revisiting Race in a Genomic Age*, ed. Barbara A. Koenig, Sandra Soo-Jin Lee, and Sarah S. Richardson (New Brunswick: Rutgers University Press, 2008), 48–49.

12. Mark D. Shriver and Rick A. Kittles, "Genetic Ancestry and the Search for Personalized Genetic Histories," in *Revisiting Race in a Genomic Age*, ed. Barbara A. Koenig, Sandra Soo-Jin Lee, and Sarah S. Richardson (New Brunswick: Rutgers University Press, 2008), 207–8.

13. Deborah A. Bolnick, "Individual Ancestry Inference and the Reification of Race as a Biological Phenomenon," in *Revisiting Race in a Genomic Age*, ed. Barbara A. Koenig, Sandra Soo-Jin Lee, and Sarah S. Richardson (New Brunswick: Rutgers University Press, 2008).

14. Patrick J. Geary, *The Myth of Nations: The Medieval Origins of Europe* (Princeton: Princeton University Press, 2001), 11.

15. Ibid., 40.

16. Neil Gerlach et al., *Becoming Biosubjects: Bodies, Systems, Technologies* (Toronto: University of Toronto Press, 2011), 9.

17. Ibid., 6.

18. Frances Watkins Harper, *Iola Leroy; Or, Shadows Uplifted*, The Schomburg Library of Nineteenth-Century Black Women Writers (Oxford: Oxford University Press, 1990), 235.

19. See also Anne-Elizabeth Murdy, *Teach the Nation: Pedagogies of Racial Uplift in U.S. Women's Writing of the 1890s* (London: Routledge, 2002), 84–85.

20. Hazel Carby, *Reconstructing Womanhood: The Emergence of the Afro-American Woman Novelist* (Oxford: Oxford University Press, 1989), 89.

21. Alexis Everett Frye, *Grammar School Geography* (Boston: Ginn, 1902).

22. Ruth Ozeki, *My Year of Meats* (New York: Penguin Books, 1998), 149–50.

23. Ibid., 150.

24. Ibid., 149.

25. Ibid., 151.

26. Ibid., 15.

27. Ibid., 152.

28. Ibid.

29. I discuss this elsewhere as a strategy of Susan Choi's. See Patricia E. Chu, "The Trials of the Ethnic Novel: Susan Choi's *American Woman* and the Post-Affirmative Action Era," *American Literary History* 23, no. 3 (2011): 529–54.

30. Ozeki, *My Year of Meats*, 11.

31. But see Ronyoung Kim, *Clay Walls* (Sag Harbor, NY: Permanent Press, 1986).

32. Ozeki, *My Year of Meats*, 235.

33. Ibid., 124–25.

34. Henry A. Giroux, *The Terror of Neoliberalism: Authoritarianism and the Eclipse of Democracy* (Boulder, CO: Paradigm, 2004), xii; Lester Spence, *Knocking the Hustle: Against the Neoliberal Turn in Black Politics* (Goleta, CA: Punctum Books, 2015); Susan George, "Neoliberalism: Nothing Owed to the Losers," *Aisling Magazine*, Tús Millennium 2000, http://www

.aislingmagazine.com/aislingmagazine/articles/TAM26/Neoliberalism.html.
  35. As an example, Cooper notes human blood no longer being treated as a national reserve outside the market. *Life as Surplus*, 9.
  36. Cooper, *Life as Surplus*, 33.
  37. Rachel C. Lee, *The Exquisite Corpse of Asian America: Biopolitics, Biosociality, and Posthuman Ecologies* (New York: New York University Press, 2014), 65.
  38. See Cheryl J. Fish, "The Toxic Body Politic: Ethnicity, Gender, and Corrective Eco-Justice in Ruth Ozeki's *My Year of Meats* and Judith Helfand and Daniel Gold's *Blue Vinyl*," *MELUS* 34, no. 2 (Summer 2009): 43–62; Laura Ahn Williams, "Gender, Race, and an Epistemology of the Abbattoir in *My Year of Meats*," *Feminist Studies* 40, no. 2 (2014): 244–72; Emily Cheng, "Meat and the Millennium: Transnational Politics of Race and Gender in Ruth Ozeki's *My Year of Meats*," *Journal of Asian American Studies* 12, no. 2 (June 2009): 191–220; Julie Sze, "Boundaries and Border Wars: DES, Technology, and Environmental Justice,"

American Quarterly 58, no. 3 (September 2006): 791–814; and Monica Chiu, "Postnational Globalization and (En)Gendered Meat Production in Ruth L. Ozeki's *My Year of Meats*," *LIT* 12 (2001): 99–128.
  39. Ozeki, *My Year of Meats*, 154.
  40. Ibid.
  41. See Sze, "Boundaries and Border Wars," for a reading of Ozeki's novel in light of Haraway's "Cyborg Manifesto."
  42. Ozeki, *My Year of Meats*, 100.
  43. Ibid., 70, 263.
  44. Ibid., 263.
  45. Ibid., 69.
  46. Angela Y. Davis, *The Prison Industrial Complex*, MP3 (Chico, CA: AK Press, 2001); Zygmunt Bauman, *Wasted Lives: Modernity and Its Outcasts* (Oxford, UK: Polity Press, 2004); Giroux, *Terror of Neoliberalism*.
  47. Ozeki, *My Year of Meats*, 70.
  48. Ibid., 30.
  49. Richard Doyle, *On Beyond Living: Rhetorical Transformations of the Life Sciences* (Stanford: Stanford University Press, 1997), 22–23.

References

Bauman, Zygmunt. *Wasted Lives: Modernity and Its Outcasts*. Oxford, UK: Polity Press, 2004.
Bolnick, Deborah A. "Individual Ancestry Inference and the Reification of Race as a Biological Phenomenon." In *Revisiting Race in a Genomic Age*, edited by Barbara A. Koenig, Sandra Soo-Jin Lee, and Sarah S. Richardson, 70–85. New Brunswick: Rutgers University Press, 2008.
Carby, Hazel. *Reconstructing Womanhood: The Emergence of the Afro-American Woman Novelist*. Oxford: Oxford University Press, 1989.
Cheng, Emily. "Meat and the Millennium: Transnational Politics of Race and Gender in Ruth Ozeki's *My Year of Meats*." *Journal of Asian American Studies* 12, no. 2 (June 2009): 191–220.
Chiu, Monica. "Postnational Globalization and (En)Gendered Meat Production in Ruth L. Ozeki's *My Year of Meats*." *LIT* 12 (2001): 99–128.
Chu, Patricia E. "The Trials of the Ethnic Novel: Susan Choi's *American Woman* and the Post–Affirmative Action Era." *American Literary History* 23, no. 3 (2011): 529–54.
Cooper, Melinda. *Life as Surplus: Biotechnology and Capitalism in the Neoliberal Era*. Seattle: University of Washington Press, 2008.
Davis, Angela Y. *The Prison Industrial Complex*. MP3. Chico, CA: AK Press, 2001.

Davis, Lennard J., and David B. Morris. "Biocultures Manifesto." *New Literary History* 38 (2007): 411–18.

Doyle, Richard. *On Beyond Living: Rhetorical Transformations of the Life Sciences*. Stanford: Stanford University Press, 1997.

Dupré, John. "What Genes Are and Why There Are No Genes for Race." In *Revisiting Race in a Genomic Age*, edited by Barbara A. Koenig, Sandra Soo-Jin Lee, and Sarah S. Richardson, 39–55. New Brunswick: Rutgers University Press, 2008.

Fish, Cheryl J. "The Toxic Body Politic: Ethnicity, Gender, and Corrective Eco-Justice in Ruth Ozeki's *My Year of Meats* and Judith Helfand and Daniel Gold's *Blue Vinyl*." *MELUS* 34, no. 2 (Summer 2009): 43–62.

Frye, Alexis Everett. *Grammar School Geography*. Boston: Ginn, 1902.

Geary, Patrick J. *The Myth of Nations: The Medieval Origins of Europe*. Princeton: Princeton University Press, 2001.

George, Susan. "Neoliberalism: Nothing Owed to the Losers." *Aisling Magazine*, Tús Millennium 2000. http://www.aislingmagazine.com/aislingmagazine/articles/TAM26/Neoliberalism.html.

Gerlach, Neil, Sheryl Hamilton, Rebecca Sullivan, and Priscilla Walton. *Becoming Biosubjects: Bodies, Systems, Technologies*. Toronto: University of Toronto Press, 2011.

Giroux, Henry A. *The Terror of Neoliberalism: Authoritarianism and the Eclipse of Democracy*. Boulder, CO: Paradigm, 2004.

Hacking, Ian. "Genetics, Biosocial Groups, and the Future of Identity." *Daedalus* (Fall 2006): 81–95.

Harper, Frances Watkins. *Iola Leroy; Or, Shadows Uplifted*. The Schomburg Library of Nineteenth-Century Black Women Writers. Oxford: Oxford University Press, 1990.

Kim, Ronyoung. *Clay Walls*. Sag Harbor, NY: Permanent Press, 1986.

Lee, Rachel C. *The Exquisite Corpse of Asian America: Biopolitics, Biosociality, and Posthuman Ecologies*. New York: New York University Press, 2014.

Murdy, Anne-Elizabeth. *Teach the Nation: Pedagogies of Racial Uplift in U.S. Women's Writing of the 1890s*. London: Routledge, 2002.

Ozeki, Ruth. *My Year of Meats*. New York: Penguin Books, 1998.

Shriver, Mark D., and Rick A. Kittles. "Genetic Ancestry and the Search for Personalized Genetic Histories." In *Revisiting Race in a Genomic Age*, edited by Barbara A. Koenig, Sandra Soo-Jin Lee, and Sarah S. Richardson, 201–14. New Brunswick: Rutgers University Press, 2008.

Spence, Lester. *Knocking the Hustle: Against the Neoliberal Turn in Black Politics*. Goleta, CA: Punctum Books, 2015.

Stevens, Hallam, and Sarah S. Richardson. "Beyond the Genome." In *Postgenomics: Perspectives on Biology After the Genome*, 1–8. Durham: Duke University Press, 2015.

Sze, Julie. "Boundaries and Border Wars: DES, Technology, and Environmental Justice." *American Quarterly* 58, no. 3 (September 2006): 791–814.

Williams, Laura Ahn. "Gender, Race, and an Epistemology of the Abbattoir in *My Year of Meats*." *Feminist Studies* 40, no. 2 (2014): 244–72.

7.
Identification Made Visible
Photographic Evidence and Russell Williams

Jonathan Finn

On 21 October 2010, Colonel Russell Williams pleaded guilty to nearly one hundred crimes, including two murders, and was given a sentence that was unparalleled in Canadian criminal justice history.[1] He received two concurrent life sentences, two concurrent ten-year sentences, and eighty-two concurrent one-year sentences. The anomalous nature of the sentencing matched that of the crimes and their perpetrator. Prior to his arrest on 7 February 2010, Williams was a high-ranking military officer in Canada, serving as commander of Canadian Forces Base (CFB) Trenton, the nation's largest military base. During questioning on 7 February, Williams admitted to a four-year crime spree that resulted in eighty-eight charges, including two murders, two counts of forcible confinement and sexual assault, and more than eighty break-and-enters.

The arrest, confession, and eventual sentencing of Russell Williams were highly public, visual events in Canada. The heinous nature of the crimes as well as Williams's high-profile military status ensured that the case received significant media attention. ProQuest's Canadian Newsstream database yields 767 results for mentions of Williams during 2010. Within that coverage, the *Globe and Mail* carried roughly eighty stories, the *Toronto Star* seventy, and the *Montreal Gazette* sixty.[2] The case was also highly visual due to the centrality of photography in Williams's crimes. Williams visually documented his crimes, ultimately producing thousands of digital images and videos that were archived and classified in folders on his computer and on external hard drives at his home.

Borrowing from visual criminology and visual culture studies, this essay takes up the intersection of identification and visual images in the Russell Williams case. Specifically, I argue that the photographs in the case played a central role in identifying Williams as a criminal both legally and in the public domain. I further argue that the images identify a contemporary culture within which images and crime have a co-constitutive relationship. At stake are changing understandings of crime, criminality, and criminal justice as they form part of our "everyday aesthetics."[3]

A "Treasure Trove" of Visual Evidence

Williams's crime spree is said to have begun in September 2007 with his first home break-in. There is speculation by authorities and journalists that he had committed previous crimes, but in the agreed-on statement of facts presented during the hearing, Williams identified the start of the spree as September 2007. In all, he completed eighty-two break-and-enters in the Orleans neighborhood of Ottawa and the small community of Tweed, both equidistant from Toronto and Ottawa. In many cases he broke into the same house on multiple occasions, including a total of nine times at one location. In each case he would spend up to several hours in the house, almost entirely in the bedrooms of girls and women, where he would dress in their underwear, pose for photographs, and masturbate to orgasm in or on their clothes, bedding, or other material. And in each case he brought a camera with him, recording dozens of photographs of himself during his poses and masturbation. He also stole many items of underwear and lingerie, amassing well over a thousand during his four-year spree.

Williams's first violent break-ins took place in September 2009. He broke into the homes of two women, blindfolded them, bound their hands, and put them in various sexual poses for lengthy photo sessions. These crimes escalated into two rape and murder cases, the first in November 2009 and the second in January 2010. In the first case, Williams broke into the home of Corporal Marie France Comeau, who worked under Williams at CFB Trenton. The attack lasted for about two hours, and like the previous victims, Comeau was bound and blindfolded. Williams repeatedly raped Comeau and recorded the entire attack with both a still and a video camera, the latter documenting him in the process of the former. The second victim was Jessica Lloyd, who lived near Williams's cottage in Tweed. Lloyd's attack lasted longer than Comeau's, and she was moved from her own home to Williams's cottage. She was attacked and raped in both locations. And as with the Comeau murder, Williams documented the assault with still photographs and on video.[4]

As previously noted, one of the anomalous features of Williams's crime spree was the extraordinary amount of visual evidence that he collected. *Globe and Mail* reporter Timothy Appleby covered the case extensively, eventually writing a book, *A New Kind of Monster*, on the topic.[5] In the book Appleby summarizes the incredible expanse of Williams's collection:

> His total haul almost defies belief. In all, he admitted to stealing and cataloging around 1,400 pieces of clothing, nearly all of it women's lingerie, and in one raid alone he took 186 items. . . . And along with the underwear were many of the thousands of photographs he took, artfully concealed inside folders and subfolders within his computer system, together with a near-complete log of all his admitted crimes, recording the dates, the places, the nature of the offense and other details.[6]

Although Williams's crime spree and his accumulation of articles and images covered a four-year span, much of the coverage of the events, including the public dissemination of images from the case, transpired over an intense few months surrounding his October arrest and conviction. Of the entire newspaper coverage of Williams's crimes, just under 50 percent of all stories related to Williams were published in October 2010. The intensity of the coverage is a direct reflection of the way Williams was identified as the perpetrator of the crimes.

Williams had been committing break-ins since September 2007, but it was the disappearance of Jessica Lloyd in January 2010 that ultimately led police to him. Police found tire tracks in the snow outside Lloyd's house and set up a roadside check near her home in order to examine the tires on passing vehicles. Williams was stopped as part of the roadside check; his tires were photographed and ultimately matched to those at Lloyd's house. After the tires were found to match, police asked Williams to voluntarily come in for questioning. Williams did so on 7 February 2010 while wearing the same boots he had worn during the Lloyd murder. As with the tire tracks, police had also collected boot prints, which they then matched to Williams at the police station. Largely in response to this photographic evidence of tire tracks and boot prints, Williams confessed to Lloyd's murder on 7 February and then detailed his vast crime spree.

It was the photographs taken by police of Williams's tires and boots that led to his identification as the murderer of Jessica Lloyd. These photographs led to his confession, which in turn led police to Williams's treasure trove of visual material and his identification as the murderer of Marie France Comeau and the perpetrator of dozens of other house break-ins and sex-based offenses. Thus, police photographs along with Williams's own photographs stood as the legal evidence of his criminal identification. Just as important, however, was his identification in the public domain.

Recognizing the public interest in the case, prosecutors and media professionals asked the presiding judge, Justice Robert Scott, to permit electronic devices in the courtroom. The judge had previously banned all such devices but reversed that decision, noting, "It is important that the press have a position in this matter."[7] Although no visual recording was allowed, reporters were permitted to use cell phones, laptops, and other electronic devices throughout the process, resulting in a days-long live blog of the trial. The unfolding of the trial in real time not only heightened the immediacy of the proceedings but brought the public into the identification of Williams's criminality in a very substantial way.

Although no cameras were allowed into the Williams hearing, the Crown did release a small number of images from the case that were published by mainstream news agencies. Throughout the entire proceedings, media corporations and members of the public were sharing and discussing the visual evidence that had been released. Of the treasure trove of materials collected by Williams, two types of images were presented publicly. First were evidentiary police photographs of Williams's collections of girls' and women's underwear. Second were the so-called lingerie break-in, or fetish break-in, photographs of Williams dressed in girls' and women's underwear, posing for the camera in the homes he illegally entered. Due to their obviously sensational nature, these lingerie break-in images received significant public attention, but there were several other types of imagery circulating as well. Police crime scene photographs of Williams's car tire tracks and boot prints circulated, as did portrait photographs of his murder victims and images of the homes and locations where the crimes were committed. Maps and chronologies were also published that documented his crime spree in spatial and temporal formats. In all, these images produced a distinct visual culture of the Williams case, enabling the public to participate in the identification of his criminality and to literally see justice prevail.

The Body Captured

Meagan Suckling has examined the media coverage of Williams across three major Canadian newspapers: the *Globe and Mail*, the *Toronto Star*, and the *National Post*.[8] Looking at the period October–December 2010, Suckling found ninety-two articles in the papers that contained at least one image from the case, with the bulk of the imagery being presented in the days during and immediately following the October hearing. In all, she found four main themes in the images: the significant police presence in the case, Williams's gradual loss of control and his transition from colonel to criminal, photographs from Williams's past, and photographs comparing Williams with other high-profile killers. Rather than focus on the content of the images, I am interested in the larger visual program within which the identification and conviction of Williams took place.

Crime, Media, Culture

In his 1995 book *Picture Theory*, W. J. T. Mitchell identified in academic work a "visual turn" away from the linguistic-based model that had dominated for decades prior.[9] At the time, this turn was manifest most clearly in the rapid emergence and growth of programs in visual culture studies;[10] by the 1980s, however, an increasing number of scholarly fields began to focus attention on the import of vision, visuality, and visual representation in their fields of practice and in the historiographies of their disciplines. Work in science studies, as pioneered by Bruno Latour, Steve Woolgar, Michael Lynch, and others, likely stands as the most extensive and rigorous turn outside of visual culture studies,[11] but the critical, self-reflexive approach to the visual as evidenced in science studies is found in numerous fields, from sport history to cartography. Regardless of the specific disciplinary practice, a defining feature of the visual turn is the move away from treating the image as a representation of a preexisting subject toward understanding the constitutive roles of images and visual material—what Latour has famously called "inscriptions"—in the construction of knowledge.

There has been a reasonably significant amount of academic work addressing the role of the visual in social constructions of crime and criminality. And while the visualization of crime, deviance, and criminality is manifest across time periods and media forms—such as in antique vase painting, medieval church sculpture, and early modern anatomical atlases—the topic has received the most significant treatment in the history of photography. John Tagg's book *The Burden of Representation* and Allan Sekula's article "The Body and the Archive" are essential early texts in this regard.[12] Both Tagg and Sekula address the constitutive role of photography in the development of the criminal subject during the modern era, a topic also taken up in my own work and that of David Green, Suren Lalvani, and Martha Merrill Umphrey, among others.[13] Jennifer Mnookin has addressed the intersection of photography and crime through an analysis of the introduction and often uneasy adoption of the photograph as evidence in the American legal system.[14] And Richard K. Sherwin has extended this same idea to look at the co-constitutive relationship between contemporary popular culture and the use of visual evidence in law.[15]

Criminology is one of the more recent fields to make the visual turn, although work critically addressing the visual predates the recent formalization of the turn. Nicole Hahn Rafter's work is central in this regard.[16] Her work on criminal anthropology, specifically on the theories of Cesare Lombroso, perfectly illustrate the careful and rigorous attention to images that is now being called for by an emerging group of visual criminologists. As Rafter shows, images and the ability to make information visible were not of supplementary importance in the history of criminology but were an essential component in criminological theories of deviance, especially within criminal anthropology, as it identified deviant (and normal) bodies according to their visible signs.

More recently, visual criminologists have expanded on the work of Rafter and others to addresses the myriad roles of the visual in understandings of crime and criminality as well as in criminological research. Eamonn Carrabine has done much to trace the relationship of crime, criminology, and visual culture, paying specific attention to photographic images.[17] Drawing chiefly from photographic theorists and historians, including Tagg and Sekula, Carrabine's work offers a historiography of the contemporaneous development of photographic technology and the modern criminal subject.[18] Carrabine issues a call to arms to criminologists, writing that "criminology has no choice but to develop a more sophisticated understanding of the visual and confront the ways in which contemporary societies are saturated with images of crime."[19] The recently published *Routledge International Handbook of Visual Criminology*, which Carrabine edited with Michelle Brown, is evidence that this call is starting to be met.[20]

Introducing the collection of essays titled *Framing Crime: Cultural Criminology and Image*, Keith Hayward stresses the particularly important role of the visual in twenty-first-century society. He summarizes: "While the everyday experience of life in contemporary Western society may or may not be suffused with crime, it is most certainly suffused with images and increasingly images of crime."[21] Importantly, Hayward stresses, like Carrabine, that such a visual turn in criminology must move beyond simply incorporating images into already established modes of analysis and instead seek to "understand and identify the various ways in which mediated processes of visual production and cultural exchange now 'constitute' the experience of crime, self, and society under conditions of late modernity."[22] As such, criminologists are increasingly turning to the role of photographs, social media, television, and other media forms to examine the constitutive roles of images in crime, criminality, and criminal justice.

The role of the visual in the Williams case exemplifies perfectly the arguments put forth by Hayward, Carrabine, Brown, and other visual criminologists about the centrality of the image in contemporary understandings of crime and criminality. It also exemplifies their insistence that criminologists move beyond addressing images purely as forms of representation. In the most immediate sense, the images taken by Williams served as direct representations of his criminality: they were taken of him by him during the commission of criminal offenses. But when we move from Williams's private archives to the more public arena of his arrest and sentencing, the images take on further significance. In a visual culture intertwined with crime and criminal justice, the Williams photographs played a key role in his identification as a criminal and in visually depicting the victory of justice over crime.

The before-and-after trope featured prominently in media coverage of the Williams case. Typically, these would show Williams as colonel (sometimes with high-ranking military and political figures) alongside images of Williams's lingerie break-ins. These helped to chronicle and promote the notion of Williams's double identity and his slide from military commander to sexual deviant. Importantly, members of the public also participated in this aspect of visualization, combining their own pictures and commentary on mainstream news sites and on social media and photo-sharing sites. The before-and-after trope was complemented by other images from the case, including courtroom sketches and police crime scene photos, all of which located Williams within the standard visual vocabulary of crime and criminal justice. It is interesting to note that the most traditional symbol of criminality—the mug shot—was absent in the media portrayal of the Williams case.

Just as the visual culture of the Williams case helped the public to literally and figuratively construct his criminality, it also helped them make sense of the horrific events involved in the crimes. Of the visual materials that performed this role, one took on somewhat mythological proportions in the public domain: the videotaped confession of Williams on 7 February 2010. Williams was brought into an Ottawa, Ontario, police station and questioned on that date for approximately nine hours. His confession came near the halfway point, after which Williams took the next few hours to detail his crimes. The public display of this confession was intense; it featured regularly on news sites, is available on YouTube and on a Williams Wikipedia page, and was published in newspapers via frame grabs.[23] The detective involved, OPP Detective-Sergeant Jim Smyth, was described as something of a hero in media accounts. The discourse around the confession suggested that this was a piece of police work par excellence and was a real-life CSI moment in which the supervillain is undone by the superhero.

The heinous and anomalous nature of Williams's crimes was made visible through the release of images of his lingerie break-ins. This resulted in a visual paradox of Williams's double identity as colonel and criminal, exemplified in the before-and-after images of him. The video confession helped provide public closure to the case by offering a familiar story of justice served. Williams's criminality and Smyth's status as superhero cop were mutually constituted in the video confession and in the ensuing public discourse surrounding the video. This is aptly illustrated in the closing line of Timson's piece for the *Globe and Mail*, in which she writes, "This is why I particularly liked the interrogation video excerpts. They showed Russell Williams stripped of all his power."[24] Timson's comment effectively summarizes the restorative function of the Williams images, especially the video confession, as they literally showed the victory of justice over crime.

Ubiquitous Photography and Everyday Aesthetics

In her now famous treatment of the Abu Ghraib images, Susan Sontag argues that "the photographs are us," a concept she uses to refute the claim made by the US administration at the time that the images were the products of a few "bad apples."[25] Instead, Sontag argues that the images are the logical outcome of a broader American culture that is unapologetically brutal and driven by a desire to capture and be captured in images. This latter point is concisely stated in her remark about one particularly troubling image depicting a group of naked prisoners stacked on one another to form a human pyramid. She writes, "There would be something missing if, after stacking the naked men, you couldn't take a picture of them."[26] In other words, for Sontag, the very act of stacking the naked prisoners was done—at least in part—to be photographed.

As part of her analysis of the Abu Ghraib images, Sontag draws parallels to lynching photographs but notes a key distinction: the former were not taken as trophies as the latter photographs were but as objects to be shared and disseminated in a world dominated by digital communication.[27] This part of Sontag's argument echoes a larger academic discourse from the 1990s through the 2000s addressing the shift from traditional analog image-making to digital formats and the resultant (digital) visual culture. Whereas earlier work by W. J. T. Mitchell, Geoffrey Batchen, and Lev Manovich examined claims related to the death of photography or the transition to digital imaging, more recent work by Paul Frosh, David Machin, Jose Van Dijck, Susan Murray, Martin Lister, and Daniel Rubenstein and Katrina Sluis extends the debate to stress the increasing prevalence of photographies in day-to-day life.[28] Such is the case with Sontag's reading of the Abu Ghraib images—they are the product of a culture fully consumed by the desire to take, send, and receive photographs, and with the advanced communication technologies to make such practices nearly limitless. Importantly, this desire is manifest across all manner of activities, from taking selfies and documenting the food on one's plate to filming clandestine scenes of police brutality and capturing the torture of human beings.

Martin Hand uses the term "ubiquitous photography" to identify and explore this changing visual landscape, with particular attention on the increasing number of cameras inserted into daily life and daily objects.[29] He stresses that this is not just a shift in modes of visual representation but a broader cultural shift, writing, "In embracing the term 'ubiquitous,' then, I am not referring simply to images: I suggest that the discourses, technologies and practices of photography have become *radically pervasive* across all domains of contemporary society."[30] And so,

The weaving of *photographies*—as images and ideas, as devices and techniques, and as practices—into every corner of contemporary society and culture produces quite a different scenario from that envisaged during the late twentieth century. Where many once imagined a future of digital simulation and virtual reality, we now arguably have the opposite: the visual publicization of ordinary life in a ubiquitous photoscape.[31]

Ubiquitous photography, then, refers not just to the proliferation of image-making hardware and software in contemporary culture but also to an ontological shift in which photography is increasingly understood as part of our every day. Susan Murray makes a similar argument, outlining the emergence of an "everyday aesthetic" through the proliferation of photo-sharing sites such as Flickr.[32] She writes, "On these sites, photography has become less about the special or rarefied moments of domestic/family living (for such things as holidays, gatherings, [and] baby photos) and more about an immediate, rather fleeting display of one's discovery of the small and mundane (such as bottles, cupcakes, trees, debris, and architectural elements)."[33]

Similarly, I want to suggest that the Williams images "are us" and that they can be read as part of our "everyday aesthetics." The images are not the product of a single man but instead identify a larger culture. Most immediately the images identify a culture with systemic problems of gendered discrimination and violence. Such problems are well documented within the Canadian military.[34] At least two large studies of the Canadian military were commissioned following the Williams case, and both cited systemic problems of gendered and sexual discrimination in the military.[35] As summarized by one of the studies, "There is an underlying sexualized culture in the CAF [Canadian Armed Forces] that is hostile to women and LGTBQ members, and conducive to more serious incidents of sexual harassment and assault."[36] Despite these studies and the many new incidents and concerns raised since 2010, gendered and sexual violence and discrimination within and outside the military remain a pressing social issue.

The Williams photographs are us in that they also identify a culture underlain with a desire, if not compulsion, to photographically record our world and our actions in it. Just as stacking the naked men in the Abu Ghraib prison was done in part to be photographed, so too were Williams's crimes committed to be photographed. Williams carried a Sony DLSR with him as an essential piece of his crime kit.[37] And we know that Williams staged and posed himself, his victims, and their belongings in order to be visually recorded. Williams's crimes and their visual representation are inextricably bound.

Finally, the Williams images are us because they identify a culture in which crime, criminality, and criminal justice are increasingly communicated, understood, and defined in visual form. Traditional police photography (of tire tracks and boot prints) initiated a sequence of events that led to the identification of Williams's visual archives, his visually recorded confession, and the hearing and sentencing, which involved the juridical and journalistic use of images. To use Hayward's term, the Williams case—from the commission of crimes to postsentencing—was "suffused" with images.

If it is clear that the Williams photos are us, less obvious is how the imagery can be read as part of an everyday aesthetic. After all, Murray, Rubenstein and Sluis, and Hand focus on the "new" forms of photography and photographic practice that are aligned with social media, and Williams did not create his images to be shared on Instagram, Flickr, or another platform. Yet there is still something very ordinary about the visual culture of the case. Police images of footprints, tire tracks, and crime scene tape are ubiquitous in television dramas, Hollywood films, and other media formats. Such images are as routine as the selfies or photographs of food that make up the subject matter of the previously noted scholars. Even the lingerie break-in photographs fit within an everyday aesthetic. These images are paradoxical: although they are clearly remarkable in terms of what we *know* they represent, their content is strangely mundane. No violence is depicted, and there are no signs of criminal activity. Instead, the photographs speak to Murray's everyday aesthetics in that they are filled with the accoutrements of bedrooms and the domestic sphere: underwear, teddy bears, flowery duvets, posters, and louvered doors. Perhaps more than anything else, it is the (accepted) banality of Williams's images that is most unsettling.

Conclusion

The thousands of images taken by Williams during his crimes, the select few displayed later in the mainstream media, and the visual materials produced by police, media personnel, and citizens constituted a distinct visual culture surrounding the Williams case. These images were more than passive representations; they played a fundamental role in the identification of Williams as a criminal both legally and in the public domain. The images produced by police, such as those of Williams's tires and boot prints, were essential in identifying Williams as a suspect and bringing him to the police station. Those same photos and others were used in the subsequent questioning of Williams captured in the now infamous video confession. That confession led to police searching Williams's homes, ultimately finding his treasure trove of photographs, videos, and stolen underwear, which was in turn central to the police and prosecution in Williams's conviction and sentencing. And as they were disseminated in the media, the images enabled the public to participate in Williams's identification as a criminal and to literally see justice prevail over crime.

On seeing photographs at the age of twelve of the concentration camps at Bergen-Belsen and Dachau, Sontag notes that she was forever changed by their immediate and long-term impact. She writes, "Once one has seen such images, one has started down the road of seeing more—and more. Images transfix. Images anesthetize." She continues, noting that though photographs inevitably make an event seem more "real," "after repeated exposure to images it also becomes less real."[38] This, I argue, is a central concern with the photographs in the Russell Williams case. They identify a contemporary visual culture suffused with images of crime and criminal justice. Within such a space, photographs of human violence, natural disasters, and police brutality circulate equally alongside selfies and photographs of birthday parties and meals. Of significant concern is the extent to which this renders images such as those of the Williams case an unremarkable part of our everyday aesthetics.

Notes

1. As a scholar of visual communication and visual culture, I stress the importance of articulating arguments in both visual and textual form. Nonetheless I have decided to refrain from publishing images in this case because of the nature of the crimes committed and the role of the photograph in those crimes. Many of the images are easily and freely available through an internet search and/or through the hyperlinks included with some of the sources here.

2. Research for this article includes a cursory search of Canadian and international newspaper stories as well as social media sources, including YouTube. Almost all the research took place after Williams's sentencing, though I did follow the case closely during 2010 due to my scholarly interest in crime and visual culture. For

brevity's sake, notes to the essay will include only sources directly cited.

3. The term comes from Susan Murray and will be discussed later in the essay. "Digital Images, Photo-Sharing, and Our Shifting Notions of Everyday Aesthetics," *Journal of Visual Culture* 7, no. 2 (August 2008): 147–63, https://doi.org/10.1177/1470412908091935.

4. A concise timeline of events is available here: "Col. Russell Williams Timeline," CBC/Radio-Canada, updated 26 September 2020, https://www.cbc.ca/news/canada/col-russell-williams-timeline-1.913312.

5. Timothy Appleby and Greg McArthur, "Evidence Exposes Scope of Depravity; Former Commander's Own Photos and Files Document Dozens of Sex Crimes," *Globe and Mail*, 19 October 2010, A7; Timothy Appleby, "The Tire Track That Broke the Case Wide Open," *Globe and Mail*, 10 February 2010, A9; Appleby, "Arrest 'a Body Blow,' General Says; Natynczyk Tells Troops to Move Forward and Wear Their Uniforms Proudly," *Globe and Mail*, 11 February 2010, A8; Appleby, "Star Lawyer Edelson to Take Colonel's Case; Veteran of Ottawa Mayor's Corruption Trial Set to Defend Williams as He Faces Pair of First-Degree Murder Charges," *Globe and Mail*, 18 February 2010, A7; Appleby, "Russell Williams Makes Court Appearance; Murder Trial Put Over till March 25," *Globe and Mail*, 19 February 2010, A4; Appleby, "Accused Commander on a Hunger Strike, but Without Demands," *Globe and Mail*, 10 April 2010, A7; Appleby, "Plea Agreement in the Works in Williams Case; 82 New Burglary-Related Charges Laid Against Ex-Colonel," *Globe and Mail*, 30 April 2010, A6; Appleby, "Calm Williams Appears via Video; 'Understood, Thank You,' Accused Says as Murder Case Adjourned for Another Month," *Globe and Mail*, 25 June 2010, A6; Appleby, "Accused Colonel Waives Right to Preliminary Hearing; Goes Directly to Trial," *Globe and Mail*, 27 August 2010, A5; Appleby, "Grief, Contempt as Families Face Williams; 'The Love in Our Family Flows Strong and Deep. He Will Never Be Able to Take That from Us,' Victim's Aunt Tells Court," *Globe and Mail*, 21 October 2010, A4; Appleby, "Williams Begins Serving Life Sentences; Former Military Commander, Called One of the Most Despicable Killers in Canadian History, Tells Court He's 'Indescribably Ashamed,'" *Globe and Mail*, 22 October 2010, A7; Appleby, *A New Kind of Monster: The Secret Life and Chilling Crimes of Colonel Russell Williams* (Toronto: Vintage Canada, 2012); Colin Freeze and Timothy Appleby, "Ex-Commander to Appear in Court April 29; Lawyer to Digest 'Substantial' Disclosure in Bizarre Case of Murders, Sex Assaults," *Globe and Mail*, 26 March 2010, A5.

6. Appleby, *New Kind of Monster*, 106.

7. "Judge Sets Rules for Media in Col. Williams Case," CTV.ca, October 14, 2010, https://www.ctvnews.ca/judge-sets-rules-for-media-in-col-williams-case-1.562962.

8. Meagan Suckling, "From Colonel to Criminal: A Visual Examination of the Construction of Russell Williams' Criminality in the News" (major research paper, Wilfrid Laurier University, 2012).

9. W. J. T. Mitchell, *Picture Theory: Essays on Verbal and Visual Representation* (Chicago: University of Chicago Press, 1995). See also Martin Jay, "That Visual Turn," *Journal of Visual Culture* 1, no. 1 (April 2002): 87–92.

10. Margaret Dikovitskaya, *Visual Culture: The Study of the Visual After the Cultural Turn* (Cambridge, MA: MIT Press, 2006).

11. Bruno Latour, *Science in Action: How to Follow Scientists and Engineers Through Society* (Cambridge, MA: Harvard University Press, 1987); Michael Lynch and Steve Woolgar, eds., *Representation in Scientific Practice* (Cambridge, MA: MIT Press, 1990).

12. John Tagg, *The Burden of Representation: Essays on Photographies and Histories* (New York: Palgrave Macmillan, 2002); Allan Sekula, "The Body and the Archive," *October* 39 (1986): 3.

13. Jonathan Finn, *Capturing the Criminal Image: From Mug Shot to Surveillance*

Society (Minneapolis: University of Minnesota Press, 2009); David Green, "On Foucault: Disciplinary Power and Photography," *Camerawork* 32 (1985): 6–9; Green, "Veins of Resemblance: Photography and Eugenics," *Oxford Art Journal* 7, no. 2 (1985): 3–16; Suren Lalvani, *Photography, Vision, and the Production of Modern Bodies* (Albany: State University of New York Press, 1996); Martha Merrill Umphrey, "The Sun Has Been Too Quick for Them: Criminal Portraiture and the Police in the Late Nineteenth Century," *Studies in Law, Politics and Society* 16 (1997): 139–63.

14. Jennifer Mnookin, "The Image of Truth: Photographic Evidence and the Power of Analogy," *Yale Journal of Law and the Humanities* 10, no. 1 (1998): 1–74.

15. Richard K. Sherwin, "Visual Literacy in Action: 'Law in the Age of Images,'" in *Visual Literacy*, ed. James Elkins (New York: Routledge, 2007), 179–94.

16. Nicole Hahn Rafter, *The Origins of Criminology: A Reader* (New York: Routledge-Cavendish, 2009); Cesare Lombroso, *Criminal Man*, trans. Nicole Hahn Rafter and Mary Gibson (Durham: Duke University Press, 2006); Cesare Lombroso and Guglielmo Ferrero, *Criminal Woman, the Prostitute, and the Normal Woman*, trans. Nicole Hahn Rafter and Mary Gibson (Durham: Duke University Press, 2003). See also David Horn, *The Criminal Body: Lombroso and the Anatomy of Deviance* (New York: Routledge, 2003).

17. Eammon Carrabine, "Just Images: Aesthetics, Ethics and Visual Criminology," *British Journal of Criminology* 52, no. 3 (1 May 2012): 463–89, https://doi.org/10.1093/bjc/azr089; Carrabine, "Images of Torture: Culture, Politics and Power," *Crime, Media, Culture: An International Journal* 7, no. 1 (April 2011): 5–30, https://doi.org/10.1177/1741659011404418; Carrabine, "Picture This: Criminology, Image and Narrative," *Crime, Media, Culture* 12, no. 2 (2016): 253–70; Carrabine, "Seeing Things: Violence, Voyeurism and the Camera," *Theoretical Criminology* 18, no. 2 (2014): 134–58.

18. Sekula, "The Body and the Archive," 3; Tagg, *Burden of Representation*.

19. Carrabine, "Just Images," 463.

20. Michelle Brown and Eamonn Carrabine, eds., *Routledge International Handbook of Visual Criminology*, Routledge International Handbooks (London: Routledge, Taylor and Francis, 2017).

21. Keith J. Hayward and Mike Presdee, eds., *Framing Crime: Cultural Criminology and the Image* (New York: Routledge, 2010), 1.

22. Hayward and Presdee, *Framing Crime*, 5.

23. "Russell Williams Confession—YouTube," accessed 26 June 2018, https://www.youtube.com/watch?v=lj7QRP37Wn0; Wikipedia, s.v. "Russell Williams (Criminal)," last modified 30 November 2020, 12:33, https://en.wikipedia.org/w/index.php?title=Russell_Williams_(criminal)&oldid=846519101.

24. Judith Timson, "The Russell Williams Onslaught: Why I Didn't Look Away; Some of My Friends Shut Their Eyes and Ears, and I Don't Blame Them. I Chose the Full Immersion Route—Because Knowledge Is Power Too," *Globe and Mail*, 22 October 2010, L2.

25. Susan Sontag, "Regarding the Torture of Others," *New York Times Magazine*, 23 May 2004, https://www.nytimes.com/2004/05/23/magazine/regarding-the-torture-of-others.html.

26. Ibid.

27. See also Dora Apel's work on lynching photographs. *Imagery of Lynching: Black Men, White Women, and the Mob* (New Brunswick: Rutgers University Press, 2004).

28. Geoffrey Batchen, *Burning with Desire: The Conception of Photography* (Cambridge, MA: MIT Press, 1997); Paul Frosh, *The Image Factory: Consumer Culture, Photography and the Visual Content Industry*, New Technologies/New Cultures (Oxford, UK: Berg, 2003); Frosh, "The Gestural Image: The Selfie, Photography Theory, and Kinesthetic Sociability," *International Journal of Communication* 9 (2015): 22; Martin Lister, "A Sack in the Sand: Photography

in the Age of Information," *Convergence: The International Journal of Research into New Media Technologies* 13, no. 3 (1 August 2007): 251–74, https://doi.org/10.1177/1354856507079176; David Machin, "Building the World's Visual Language: The Increasing Global Importance of Image Banks in Corporate Media," *Visual Communication* 3, no. 3 (October 2004): 316–36, https://doi.org/10.1177/1470357204045785; Mitchell, *Picture Theory*; Murray, "Digital Images"; Daniel Rubinstein and Katrina Sluis, "A Life More Photographic: Mapping the Networked Image," *Photographies* 1, no. 1 (March 2008): 9–28, https://doi.org/10.1080/17540760701785842; José van Dijck, "Digital Photography: Communication, Identity, Memory," *Visual Communication* 7, no. 1 (February 2008): 57–76, https://doi.org/10.1177/1470357207084865.

29. Martin Hand, *Ubiquitous Photography* (Cambridge, UK: Polity, 2012).
30. Ibid., 12.
31. Ibid., 1.
32. Murray, "Digital Images."
33. Ibid., 151.
34. Natalie Clancy, "Ottawa Expected to Halt Lawsuit by Compensating Past and Present Female RCMP Employees over Harassment Claims," CBC/Radio-Canada, updated 6 October 2016, https://www.cbc.ca/news/canada/british-columbia/rcmp-harassment-mounties-discrimination-1.3792451; Adam Cotter, *Sexual Misconduct in the Canadian Armed Forces, 2016*, Statistics Canada Catalogue no. 85-603-X (Ottawa: Statistics Canada, 28 November 2016), http://publications.gc.ca/collections/collection_2016/statcan/85-603-x2016001-eng.pdf; Marie Deschamps, *External Review into Sexual Misconduct and Sexual Harassment in the Canadian Armed Forces* (Ottawa: External Review Authority, 27 March 2015), http://www.forces.gc.ca/en/caf-community-support-services/external-review-sexual-mh-2015/summary.page; James Cudmore, "Royal Military College Head Apologizes for Cadets' Behavior at Harassment Seminars," CBC/Radio-Canada, updated 22 May 2015, https://www.cbc.ca/news/politics/royal-military-college-head-apologizes-for-cadets-behaviour-at-harassment-seminars-1.3082348; Lee Berthiaume, "Concerns Prompt Sweeping Review of Royal Military College," *The Star*, 2 November 2016, https://www.thestar.com/news/canada/2016/11/02/concerns-prompt-sweeping-review-of-royal-military-college.html; Bruce Campion-Smith, "Nearly 1,000 Canadian Soldiers Report Being Sexually Assaulted over Past Year," *The Star*, 28 November 2016, https://www.thestar.com/news/canada/2016/11/28/960-canadian-soldiers-report-being-sexually-assaulted-last-year.html.

35. Cotter, *Sexual Misconduct*; Deschamps, *External Review into Sexual Misconduct*.
36. Deschamps, *External Review into Sexual Misconduct*, ii.
37. Appleby, *New Kind of Monster*.
38. Susan Sontag, *On Photography* (New York: Doubleday, 1977), 20.

References

Apel, Dora. *Imagery of Lynching: Black Men, White Women, and the Mob*. New Brunswick: Rutgers University Press, 2004.

Appleby, Timothy. "Accused Colonel Waives Right to Preliminary Hearing; Goes Directly to Trial." *Globe and Mail*, 27 August 2010, A5.

———. "Accused Commander on a Hunger Strike, but Without Demands." *Globe and Mail*, 10 April 2010, A7.

———. "Arrest 'a Body Blow,' General Says; Natynczyk Tells Troops to Move Forward and Wear Their Uniforms Proudly." *Globe and Mail*, 11 February 2010, A8.

———. "Calm Williams Appears via Video; 'Understood, Thank You,' Accused Says as Murder Case Adjourned for Another Month." *Globe and Mail*, 25 June 2010, A6.

———. "Grief, Contempt as Families Face Williams; 'The Love in Our Family Flows Strong and Deep. He Will Never Be Able to Take That from Us,' Victim's Aunt Tells Court." *Globe and Mail*, 21 October 2010, A4.

———. *A New Kind of Monster: The Secret Life and Chilling Crimes of Colonel Russell Williams*. Toronto: Vintage Canada, 2012.

———. "Plea Agreement in the Works in Williams Case; 82 New Burglary-Related Charges Laid Against Ex-Colonel." *Globe and Mail*, 30 April 2010, A6.

———. "Russell Williams Makes Court Appearance; Murder Trial Put Over till March 25." *Globe and Mail*, 19 February 2010, A4.

———. "Star Lawyer Edelson to Take Colonel's Case; Veteran of Ottawa Mayor's Corruption Trial Set to Defend Williams as He Faces Pair of First-Degree Murder Charges." *Globe and Mail*, 18 February 2010, A7.

———. "The Tire Track That Broke the Case Wide Open." *Globe and Mail*, 10 February 2010, A9.

———. "Williams Begins Serving Life Sentences; Former Military Commander, Called One of the Most Despicable Killers in Canadian History, Tells Court He's 'Indescribably Ashamed.'" *Globe and Mail*, 22 October 2010, A7.

Appleby, Timothy, and Greg McArthur. "Evidence Exposes Scope of Depravity; Former Commander's Own Photos and Files Document Dozens of Sex Crimes." *Globe and Mail*, 19 October 2010, A7.

Batchen, Geoffrey. *Burning with Desire: The Conception of Photography*. Cambridge, MA: MIT Press, 1997.

Berthiaume, Lee. "Concerns Prompt Sweeping Review of Royal Military College." *The Star*, 2 November 2016. https://www.thestar.com/news/canada/2016/11/02/concerns-prompt-sweeping-review-of-royal-military-college.html.

Brown, Michelle, and Eamonn Carrabine, eds. *Routledge International Handbook of Visual Criminology*. Routledge International Handbooks. London: Routledge, Taylor and Francis, 2017.

Campion-Smith, Bruce. "Nearly 1,000 Canadian Soldiers Report Being Sexually Assaulted over Past Year." *The Star*, 28 November 2016. https://www.thestar.com/news/canada/2016/11/28/960-canadian-soldiers-report-being-sexually-assaulted-last-year.html.

Carrabine, Eammon. "Images of Torture: Culture, Politics and Power." *Crime, Media, Culture: An International Journal* 7, no. 1 (April 2011): 5–30. https://doi.org/10.1177/1741659011404418.

———. "Just Images: Aesthetics, Ethics and Visual Criminology." *British Journal of Criminology* 52, no. 3 (1 May 2012): 463–89. https://doi.org/10.1093/bjc/azr089.

———. "Picture This: Criminology, Image and Narrative." *Crime, Media, Culture* 12, no. 2 (2016): 253–70.

———. "Seeing Things: Violence, Voyeurism and the Camera." *Theoretical Criminology* 18, no. 2 (2014): 134–58.

CBC/Radio-Canada. "Col. Russell Williams Timeline." Updated 26 September 2020. https://www.cbc.ca/news/canada/col-russell-williams-timeline-1.913312.

Clancy, Natalie. "Ottawa Expected to Halt Lawsuit by Compensating Past and Present Female RCMP Employees over Harassment Claims." CBC/Radio-Canada. Updated 6 October 2016. https://www.cbc.ca/news/canada/british-columbia/rcmp-harassment-mounties-discrimination-1.3792451.

Cotter, Adam. *Sexual Misconduct in the Canadian Armed Forces, 2016*. Statistics Canada Catalogue no. 85-603-X. Ottawa: Statistics Canada, 28 November 2016. http://publications.gc.ca

/collections/collection_2016/statcan/85-603-x2016001-eng.pdf.

Cudmore, James. "Royal Military College Head Apologizes for Cadets' Behavior at Harassment Seminars.." CBC/Radio-Canada. Updated 22 May 2015. https://www.cbc.ca/news/politics/royal-military-college-head-apologizes-for-cadets-behaviour-at-harassment-seminars-1.3082348.

Deschamps, Marie. *External Review into Sexual Misconduct and Sexual Harassment in the Canadian Armed Forces*. Ottawa: External Review Authority, 27 March 2015. http://www.forces.gc.ca/en/caf-community-support-services/external-review-sexual-mh-2015/summary.page.

Dijck, José van. "Digital Photography: Communication, Identity, Memory." *Visual Communication* 7, no. 1 (February 2008): 57–76. https://doi.org/10.1177/1470357207084865.

Dikovitskaya, Margaret. *Visual Culture: The Study of the Visual After the Cultural Turn*. Cambridge, MA: MIT Press, 2006.

Finn, Jonathan. *Capturing the Criminal Image: From Mug Shot to Surveillance Society*. Minneapolis: University of Minnesota Press, 2009.

Freeze, Colin, and Timothy Appleby. "Ex-Commander to Appear in Court April 29; Lawyer to Digest 'Substantial' Disclosure in Bizarre Case of Murders, Sex Assaults." *Globe and Mail*, 26 March 2010, A5.

Frosh, Paul. "The Gestural Image: The Selfie, Photography Theory, and Kinesthetic Sociability." *International Journal of Communication* 9 (2015): 22.

———. *The Image Factory: Consumer Culture, Photography and the Visual Content Industry*. New Technologies/New Cultures. Oxford, UK: Berg, 2003.

Green, David. "On Foucault: Disciplinary Power and Photography." *Camerawork* 32 (1985): 6–9.

———. "Veins of Resemblance: Photography and Eugenics." *Oxford Art Journal* 7, no. 2 (1985): 3–16.

Hand, Martin. *Ubiquitous Photography*. Cambridge, UK: Polity, 2012.

Hayward, Keith J., and Mike Presdee, eds. *Framing Crime: Cultural Criminology and the Image*. New York: Routledge, 2010.

Horn, David. *The Criminal Body: Lombroso and the Anatomy of Deviance*. New York: Routledge, 2003.

Jay, Martin. "That Visual Turn." *Journal of Visual Culture* 1, no. 1 (April 2002): 87–92.

"Judge Sets Rules for Media in Col. Williams Case." CTV.ca. October 14, 2010. https://www.ctvnews.ca/judge-sets-rules-for-media-in-col-williams-case-1.562962.

Lalvani, Suren. *Photography, Vision, and the Production of Modern Bodies*. Albany: State University of New York Press, 1996.

Latour, Bruno. *Science in Action: How to Follow Scientists and Engineers Through Society*. Cambridge, MA: Harvard University Press, 1987.

Lister, Martin. "A Sack in the Sand: Photography in the Age of Information." *Convergence: The International Journal of Research into New Media Technologies* 13, no. 3 (1 August 2007): 251–74. https://doi.org/10.1177/1354856507079176.

Lombroso, Cesare. *Criminal Man*. Translated by Nicole Hahn Rafter and Mary Gibson. Raleigh: Duke University Press, 2006.

Lombroso, Cesare, and Guglielmo Ferrero. *Criminal Woman, the Prostitute, and the Normal Woman*. Translated by Nicole Hahn Rafter and Mary Gibson. Raleigh: Duke University Press, 2003.

Lynch, Michael, and Steve Woolgar, eds. *Representation in Scientific Practice*. Cambridge, MA: MIT Press, 1990.

Machin, David. "Building the World's Visual Language: The Increasing Global Importance of Image Banks in Corporate Media." *Visual Communication* 3, no. 3 (October 2004): 316–36. https://doi.org/10.1177/1470357204045785.

McArthur, Greg. "The Colonel Murdered Women but Loved His Cat; A Rare Look at Inner Life of Former Commander Pleading to 80 Crimes." *Globe and Mail*, 18 October 2010, A1.

Mitchell, W. J. T. *Picture Theory: Essays on Verbal and Visual Representation.* Chicago: University of Chicago Press, 1995.

Mnookin, Jennifer. "The Image of Truth: Photographic Evidence and the Power of Analogy." *Yale Journal of Law and the Humanities* 10, no. 1 (1998): 1–74.

Murray, Susan. "Digital Images, Photo-Sharing, and Our Shifting Notions of Everyday Aesthetics." *Journal of Visual Culture* 7, no. 2 (August 2008): 147–63. https://doi.org/10.1177/1470412908091935.

Rafter, Nicole Hahn. *The Origins of Criminology: A Reader.* New York: Routledge-Cavendish, 2009.

Rubinstein, Daniel, and Katrina Sluis. "A Life More Photographic: Mapping the Networked Image." *Photographies* 1, no. 1 (March 2008): 9–28. https://doi.org/10.1080/17540760701785842.

"Russell Williams Confession—YouTube." Accessed June 26, 2018. https://www.youtube.com/watch?v=lj7QRP37Wno.

Sekula, Allan. "The Body and the Archive." *October* 39 (1986): 3–64. https://doi.org/10.2307/778312.

Sherwin, Richard K. "Visual Literacy in Action: 'Law in the Age of Images.'" In *Visual Literacy*, edited by James Elkins, 179–94. New York: Routledge, 2007.

Sontag, Susan. *On Photography.* New York: Doubleday, 1977.

———. "Regarding the Torture of Others." *New York Times Magazine*, 23 May 2004. https://www.nytimes.com/2004/05/23/magazine/regarding-the-torture-of-others.html.

Suckling, Meagan. "From Colonel to Criminal: A Visual Examination of the Construction of Russell Williams' Criminality in the News." Major research paper, Wilfrid Laurier University, 2012.

Tagg, John. *The Burden of Representation: Essays on Photographies and Histories.* Communications and Culture. New York: Palgrave Macmillan, 2002.

Timson, Judith. "The Russell Williams Onslaught: Why I Didn't Look Away; Some of My Friends Shut Their Eyes and Ears, and I Don't Blame Them. I Chose the Full Immersion Route—Because Knowledge Is Power Too." *Globe and Mail*, 22 October 2010.

Umphrey, Martha Merrill. "The Sun Has Been Too Quick for Them: Criminal Portraiture and the Police in the Late Nineteenth Century." *Studies in Law, Politics and Society* 16 (1997): 139–63.

Part 3
Surveillant Technologies

8.
The Face in the Biometric Passport

Liv Hausken

In 2005, the International Civil Aviation Organization (ICAO) approved a new standard for international passports. They recommended including biometric information on a chip and decided that facial recognition technology (FRT) should be the globally interoperable biometric technology. All contracting states had agreed that they should begin issuing international passports conforming to these new standards no later than 2010,[1] but by 2006, sixty countries had already started issuing biometric passports.[2] This number increased rapidly, and today more than 120 countries have issued them, including most countries in Europe and North America, large parts of South America and Asia, and an ever-increasing proportion of African countries.[3] This development has happened very quickly and almost without public debate. Why has the introduction of biometrics in travel and identification documents not been widely discussed in democratic societies?

There are several possible explanations for this absence of debate, many of which are related to the political climate in the United States and Europe in the wake of 9/11. The ICAO argues that work on developing biometric passports began long before the 9/11 attacks,[4] but biometrics in passports were not a major agenda item.[5] As pointed out by Mark B. Salter and Can E. Mutlu (2011), however, even if many of these policies are "not entirely cut from new cloth," and "the dream of perfect security and smart borders had included a number of these programs, before the attacks—9/11 gave political capital to allow for the faster or wider securitization."[6] Kristrún Gunnarsdóttir and Kjetil Rommetveit examine what they consider a *policy vacuum* in the European Union after 9/11, in which legislation was prepared and put into force on biometric passports.[7] They convincingly describe a political climate of *urgency* both in the European Union and the United States in which biometrics were presented as a public good based on a simple division between "us" and "them."[8] The impression of urgency was also a result of the US Enhanced Border Security and Visa Entry Reform Act of 2002, which in praxis required biometric passports of all twenty-seven member states of the Visa Waiver Program by 2006.[9] In general, tight deadlines are likely to promote rule following.[10] According to Jonathan P. Aus, the tight deadline imposed by the United States in the Enhanced Border Security and Visa Entry Reform Act of 2002 resulted in comparatively ill-conceived EU action on biometric passports.[11] Hence, both in the United States and in the European Union, a public debate on biometric passports seems to have been treated as a foregone conclusion.[12]

I argue, however, that an important and underestimated explanation for the absence of public debate on the introduction of biometric passports may be in the choice of FRT as the only mandatory biometric identification method required for new international passports. This choice is remarkable. Although the accuracy rate for all biometric systems is gradually being improved, comparative analyses of biometric recognition modalities still show a medium-low accuracy rate for facial recognition systems in contrast to iris recognition, finger vein recognition, retinal scanning, or fingerprinting, all of which achieve a high score on technical reliability.[13] For reasons that will be discussed later, FRT has a consistently high rate of social acceptance when compared with the other alternatives.[14] It would seem that with the introduction of biometrics in international passports, the use of FRT appears to have been the path of least social and political resistance.

Despite FRT's inherently problematic lack of accuracy, my claim that the international community has perhaps too easily adopted FRT comes from my investigations of publicly available policy documents accompanying the introduction of biometric passports. In particular, I have studied the international standard for biometric passports as specified in *Doc 9303*, the ICAO document that has set the standards for machine-readable passports since 1980 and that has also specified biometric requirements since its sixth edition in 2006.[15] I will present a key part of this large study here, focusing on the statements in this policy document that support the introduction of FRTs as the primary—and only mandatory—biometrics in passport inspection systems. This is a textual analysis of ICAO's *Doc 9303*, focusing on the introduction of biometrics in general and the support of facial biometrics in particular. As such, the analysis is firmly rooted in the humanities, with a reading of the text's *meaning potential*. The document will be considered as a text that is *open for reading* rather than as one looking for the intentions of its authors or to specific actors involved in the processes that produced it. As Paul Ricoeur once stated, "The text's career escapes the finite horizon lived by its author. What the text says now matters more than what the author meant to say, and every exegesis unfolds its procedures within the circumference of a meaning that has broken its mooring to the [intention] of its author."[16] For the purposes of this analysis, the ICAO document will be treated as an *open work*, the meaning of which is suspended as it opens up new references and receives fresh relevance from them.[17] This textual analysis of *Doc 9303* is therefore not just an analysis of statements about regulations of passports and biometrics but also a cultural and social analysis of ideas and values related to faciality, identity, and identification as well as to technology, security, and media history.

The Necessity of Technical Reliability

Why introduce biometrics in passports? According to the ICAO's *Doc 9303*, including biometrics in international passports is a crucial measure for allowing authorities to connect the passport to the traveler: it is of decisive importance "to guarantee that the holder assigned a travel document (eMRTD) by the issuing State is the person at a receiving State purporting to be that same holder."[18] To guarantee this, *Doc 9303* states, "the only method of relating the person irrevocably to his travel document is to have a physiological characteristic, i.e., a biometric, of that person associated with his travel document in a tamper-proof manner."[19] It is a matter of technical reliability.

In addition to the urgency to ensure accurate verification of identity, the passport inspector should also be able to compare the traveler's identity against a watch list,[20] that is, a list of suspected criminals and other unwanted persons. While verification is the performance of a one-to-one match between traveler and document, the watch-list task is somewhat more complicated. In *Doc 9303*, this task is defined as a "hybrid of one-to-many identification and one-to-one verification" and is characterized as a "one-to-a-few process . . . comparing a submitted biometric sample against a small number of biometric reference templates on file."[21] Without getting into a debate as to whether the allegedly very high number of names on US authorities' watch lists can be characterized as "a small number," we face here what Lucas D. Introna and Helen Nissenbaum call an open-set identification task.[22] It is an *identification* task in the sense that the passport inspector needs to search for a match in a collection of templates representing everyone on the lists.[23] It is, by definition, an *open-set* task because it is not known whether the traveler is on the watch list. If there is no match, the problem is, as Introna and Nissenbaum point out, that the authorities "do not know whether the system made a mistake or whether the identity is simply not in the reference database in the first instance."[24] Thus, it is vital to develop a technical inspection system in which the recognition rate is as high as possible and the false alarm rate is, correspondingly, as low as possible.

The watch-list task represents an additional challenge in that the data in the reference database, which includes biometric data and other information, may not have been collected under controlled conditions.[25] This is less of a problem for the verification task, assuming that the passport application process controls the environment so as to ensure high-quality data. The ICAO "vision for the application of biometrics technology" therefore includes not just specification of a primary interoperable form of biometrics technology for use in border control (verification, watch lists) but also specification of the biometrics technologies for use by document issuers (identification, verification, and watch lists).[26] The quality of a biometric system is never better than the quality of the data enrolled in the system.

The question of the use of biometrics in passports is thus not just a matter of verification of identity, as is often argued. It is also a question of identification, as it improves the quality of the background checks performed as part of the passport application process[27] and of the searches to identify a person or document against a list of "wanted" individuals. To fulfill these requirements, *Doc 9303* stresses the need for technical reliability.

Against this backdrop, ICAO's choice of FRT as the primary—and the only mandatory—biometric identification method for international passports seems peculiar. As underlined by experts in the field, computerized facial recognition is a difficult technology to engineer. Facial traits vary over time, and even a slight difference in expression can affect recognition. Kelly A. Gates stresses how FRT represents "an especially challenging technical problem and puts it at a disadvantage relative to other biometrics in terms of its level of development, ease of adoption and use, and general viability."[28] In his *Guide to Biometrics for Large-Scale Systems*, Julian Ashbourn underlines that FRT does not match the accuracy provided by certain other techniques, such as fingerprinting and iris recognition.[29] It is characterized as "weak biometrics," whereas the recognition of fingerprints, irises, and hand veins are categorized as "strong biometrics."[30] Most studies also confirm that face recognition systems are highly dependent on camera distance, angle, and lighting. And even if all these factors are controlled, comparative studies of biometric modalities have highlighted that "it is questionable whether the face itself, without any contextual information, is a sufficient basis for recognizing a person from a large number of identities with an extremely high level of confidence."[31]

The choice of FRT as the primary biometric modality in biometric passports may seem even more striking considering that international border controls have been implementing unmanned gates with biometric readers during exactly the same time period. Although there is human backup at these gates, one might think it would be of great importance to safety and efficiency[32] to have a technically reliable system in place. So why is FRT recommended as the primary—and the only mandatory—biometric for global interoperability in passport inspection systems?

Why the Face?

Doc 9303 neither compares biometric modalities nor gives an account of what discussions have been held in the five-year period the ICAO refers to as the "preparatory work" before making the choice to recommend facial recognition as the globally interoperable biometric technology for passports. *Doc 9303* only presents a list of statements supporting their choice. "In reaching this conclusion," *Doc 9303* asserts, "ICAO observed that for the majority of States[,] the following advantages [are] applied to facial images." They then render a list of eleven points,[33] the analysis of which will be the main focus of this chapter.

1. Facial photographs do not disclose information that the person does not routinely disclose to the general public.
2. The photograph (facial image) is already socially and culturally accepted internationally.
3. The facial image is already collected and verified routinely as part of the eMRTD application form process in order to produce an eMRTD to Doc 9303 specifications.
4. The public is already aware of the capture of a facial image and its use for identity verification purposes.

5. The capture of a facial image is non-intrusive. The end user does not have to touch or interact with a physical device for a substantial timeframe to be enrolled.
6. Facial image capture does not require new and costly enrollment procedures to be introduced.
7. Capture of a facial image can be deployed relatively immediately, and the opportunity to capture facial images retrospectively is also available.
8. Many States have a legacy database of facial images, captured as part of the digitized production of travel document photographs, which can be verified against new images for identity comparison purposes.
9. In appropriate circumstances, as decided by the issuing State, a facial image can be captured from an endorsed photograph, not requiring the person to be physically present.
10. For watch lists, a photograph of the face is generally the only biometric available for comparison.
11. Human verification of the biometric against the photograph/person is relatively simple and a familiar process for border control authorities.

The most striking thing about this list is the absence of technical arguments. None of the eleven points emphasize the technical benefits of FRT. Considering that biometrics is asserted as the "only method of relating the person irrevocably to his travel document," this must be regarded as noteworthy.[34] Instead of highlighting the technical advantages of FRTs, the list concerns pragmatic matters. One may sort the statements into two nonexclusive categories: sociocultural considerations (1, 2, 4, 5, and 11) and practical-economic benefits (3, 5, 6, 7, 8, 9, 10, and 11). The last statement, "Human verification of the biometric against the photograph/person is relatively simple and a familiar process for border control authorities," as well as the fifth on the list, "The capture of a facial image is non-intrusive. The end user does not have to touch or interact with a physical device for a substantial timeframe to be enrolled," fit equally well in both categories.

If we take the latter category first, one may briefly summarize these statements as saying that implementing FRT as mandatory biometrics in passport inspection systems requires little or no new investments; it is based on procedures and techniques already well known, and with minor adjustments, old and established technologies can be used to facilitate the new system; practically speaking, we are almost there: "The facial image is already collected and verified routinely" (3). It "does not require new and costly . . . procedures" (6). It "can be deployed relatively immediately" and it is also possible to create "facial images retrospectively" (7). Many states already have digitized passport photographs in older database technologies that easily can be implemented in the new system (8), which would not even require the person to be physically present (9); and sometimes, a photograph of the face is the "only biometric available for comparison" anyway (10). Since the "end user does not have to touch or interact with a physical device for a substantial timeframe to be enrolled" (5), one may collect real-time data from an individual without producing interruptions and delays. And, finally, there will be no need for expensive and time-consuming training of passport inspectors (11).

A similar pragmatic approach is expressed in the sociocultural considerations on the list; what is in the process of implementation is already well known and socially and culturally accepted. This is like showing your face in the streets—it's a type of personal information social beings voluntarily reveal on a regular basis (1). The facial photograph "is already socially and culturally accepted internationally" (2). "The public is already aware of the capture of a facial image and its use for identity verification purposes" (4). It is "nonintrusive" and therefore considered socially acceptable (5). It will not appear new and unaccustomed for border control authorities (11). In brief, the list seems to state that it will all be socially and culturally acceptable.

It is not in itself surprising that the introduction of biometrics for international passports brings with it practical, economic, social, and cultural considerations. On the contrary, it is entirely appropriate, both rhetorically and politically. But in addition to the startling fact that none of the statements emphasize technical accuracy, the list gives the impression that the introduction of biometrics in international passports hardly represents anything new at all. How can it then be imperative to implement? Let us take a closer look at this list of statements.

Photography as Method

If we look at this list more closely, we will discover that it is not really focused on FRT but on photographic pictures.[35] The very first point refers explicitly to "facial photographs." The second entry emphasizes, in brackets, that the photograph in question is a facial image. Hence, when the next seven statements

(3–9) use the term "facial image," it is reasonable to believe that they are still referring to facial photographs.

The last two of these statements (8–9) reintroduce the notion of photography, even though the term "facial image" still appears. The eighth entry points out that many States already have databases "of facial images, captured as part of the digitized production of travel document photographs" (8). In the next statement, it is claimed that "a facial image can be captured from an endorsed photograph" (9). Thus, the facial image appears first to be *synonymous* with facial photography (2–7) before emerging as a possible *part* of a process of producing passport photographs (8) or being produced from photographs (9). The reintroduction of the notion of photography points back to the first two statements, as if to confirm that we are still talking about photographic pictures. By the same token, photography acts as a bridge to the final statements on the list.

The notion of the photographic picture is also persistently present in the list's last two statements (10–11). Here, the term "facial image" does not apply. Instead, the notion of biometrics is introduced for the very first time on this list. This introduction produces another ambiguous distinction between two concepts, this time between photography and biometrics; the tenth entry treats a photograph of the face as though it were *already* biometric: "For watch lists, a photograph of the face is generally the only biometric available for comparison" (10). The facial photograph is thus characterized as a biometric identifier. In the last point of the list, however, "the biometric" is verified "against the photograph/person" (11), indicating that photographs and biometrics are there to be compared.

Arguably, the attention paid to photography in the opening statements (1–2) and in the latter third of the list (8–11) creates the impression that photography works as a frame for all the statements to which *Doc 9303* refers as supporting FRT in passports. The two opening statements offer the facial photograph as a foothold in the field before continuing the reading of the rest of the list. Hence, the list demonstrates a sliding movement from the facial photograph to facial biometrics, via the facial image, without this elision being explicitly mentioned or discussed. In reading the list from beginning to end, it is as though the text incorporates the notions of facial image and biometry just in passing. It is an almost imperceptible transformation from the familiar to the unknown, without throwing the familiar away during the journey. If this reading is seen as a process, FRT may appear to be legitimized based on a naturalized technology (photography), which is generally accepted for use in contexts involving IDs (as underlined in the fourth entry). As Jane Caplan convincingly argues, any standardized system of identification and recognition depends for its effectiveness both on the stability or replicability of its operations and on their relative invisibility.³⁶

A demonstration of photography as a naturalized technology is underlined with all possible clarity in the very first statement on the list: "Facial photographs do not disclose information that the person does not routinely disclose to the general public." This phrase conflates "the person" with "the face," and "the face" with the "facial photograph." It explicitly claims that a photo of someone's face does not provide more or other information than the person already discloses publicly and on a regular basis. This statement rhetorically transforms the identity of the person into a piece of information, a transformation that occurs on the basis of the dual function of ID photography as self-presentation and object of identification. The photographic portrait on display in the passport is not just there to identify the traveler but also the other way around—to confirm that the traveler is the rightful owner of the document and hence respected as such and free to travel. This very first statement on the list seems to insist on this familiarity with photography and suggests that respected citizens will get their ID photos approved and their respect confirmed. Reading through the list from beginning to end seems to rhetorically maintain this familiarity, trust, and respect and to help transfer this trust and respect to the new biometric technology toward the end of the list.

To briefly sum up so far, the presentation of statements in *Doc 9303* supporting FRT in passport inspection systems seems to demonstrate that facial biometrics represent a practical, economic, and politically easy way to introduce automated biometrics in passports. This is due to the sense of a close relation between face and identity (and identification), via the naturalized technology of photography, and to an attempt (intended or not) to transfer that familiarity with photography to FRT. In reality, this is a complex conflation of the conceptions of face, photography, and FRT.

This attempt to transfer trust from photography to FRT is partly accomplished by the choice of words and the sliding transitions between concepts. But the order of the statements also seems to play a significant role. As underlined by Paul Ricoeur, "a text is more than a linear succession of sentences. It is a cumulative, holistic process. This specific structure of the text cannot be derived from that of the sentence. Therefore, the kind of 'plurivocity' which belongs to texts as texts is something other than the polysemy of individual words in ordinary language and the ambiguity of individual sentences. This plurivocity is typical of the text considered as a whole, open to several readings and to several constructions."[37] The numbered list from 2006 could be read as a preferred order, a recommended sequence, possibly even a hierarchy. The bullet points from the 2015 version of the list seem to neutralize such guidelines, however, freeing the different statements from the set sequence so that they can be read in any order. Reading the list as a sequence from top to bottom seems to demonstrate that given a familiarity with photography, nothing is actually new with the introduction of facial biometrics in passports. But what does this text say if we read it as a sequence from the bottom up?

Reading Against the Grain

To introduce this notion of reading against the grain is not to imply that the earlier analysis represents a reading *with* the grain, sympathetic to the author's intention and willing to accept the text's values and beliefs. A close, critical reading of a text is always to some extent reading against the grain— scrutinizing the text, drawing attention to gaps and to ideas that seem to be taken for granted. The experiment to be undertaken in what follows will be a reading against the grain in a very literal sense. Earlier I read the seemingly neutral list as a sequence from top to bottom. I will now investigate a different order and thus create an alternative sequence. Since the notion of biometrics is introduced in the last two statements, it may be fruitful to start there, at the end, and see how the text presents itself when we start reading from the point at which the introduction of biometrics is explicit.

When we read the list from the bottom up, there is no doubt that this first statement confirms the support of biometrics. It asserts that "human verification of the biometric against the photograph/person is relatively simple and a familiar process for border control authorities" (11). It is not entirely clear, though, what is being compared in the "verification of the biometric against the photograph/person" or what makes this comparison "human." As "verification" is the performance of a one-to-one match between traveler and document, "the biometric" could refer to data stored on the chip (in the document), to be compared with a new biometric sample (of the traveler) created by way of a camera or scanning device at the border inspection. This will harmonize with the definition of "verification" in the *Doc 9303* glossary, which is "the process of comparing a submitted biometric sample against the biometric reference template of a single enrollee whose identity is being claimed, to determine whether it matches the enrollee's template."[38] Strictly speaking, then, "the biometric" is not compared to the "person" but to another set of biometrics. As pointed out by Salter and Mutlu (2011), "Biometrics . . . cannot assert a connection between a dossier—a body—and an identifier (a person, a narrative, a story)—biometrics can merely freeze the moment of congruity between different forms of identity—corporeal, bureaucratic, social."[39] Robin Cooper Feldman stresses that "biometrics are merely a form of data."[40] Hence, as Ashbourn has underlined, "all we are doing is comparing two sets of [biometric] data and, according to predefined criteria, reaching a conclusion as to whether they are alike enough to be considered a match."[41] It is, in other words, a question of "biometric matching," which the glossary of *Doc 9303* much more explicitly explains as "the process of using an algorithm that compares templates derived from the biometric reference and from the live biometric input, resulting in a determination of match or non-match."[42] Although it may be people involved in such a comparison of two biometric data sets, one may wonder that such an assessment is characterized as "human verification" and not simply as "biometric verification."[43]

It is also not clear what is compared when verifying "the biometric against the photograph." But if we assume that "the biometric," in line with the earlier interpretation, refers to the biometric data on the chip, it seems reasonable to consider that the task is to control the correlation between the chip and the photographic picture on display at the passport's ID page. The purpose of this could be to investigate whether the document has been tampered with. Conceptually, then, "the biometric" here refers to photographically generated biometric data stored on a passport chip, while photography is a notion reserved for the visual picture. One could imagine that the "human" part of this verification process could refer to the human assistance of an automated system (or the human backup, in cases of unmanned gates). It is of course also possible to visually compare the face of the traveler and the photograph on display in the passport, but this is not in and of itself a "verification of the biometric" as understood in *Doc 9303*. Rather, it is a human verification of the photograph against the appearance of the person as part of a manual check of travel documents. In light of all this, it seems reasonable to argue that this first statement in our reading from the bottom up seeks to establish the checkpoint situation as human and therefore well known and to characterize biometrics as something object-like, something on display, comparable to other items rather than to a complex operation, in which the well-known (faces and photographs) are considered input in a far less known and quite large-scale automated system.⁴⁴

The tenth statement, and the second in this alternative sequence, introduces the watch list and, with it, the second task for biometrics in passport inspection systems: identification. It states, "For watch lists, a photograph of the face is generally the only biometric available for comparison." The photograph in question is most likely not the portrait on display in the traveler's passport but rather a picture in a reference database. This photograph is characterized as (a) biometric. The message seems to be that to perform this task in pursuit of criminals and other undesirable individuals, it would not be advisable to decide on another biometric modality, such as fingerprinting or iris scanning, because such data are normally not available. The photograph has gone from being comparable with, and hence different from, biometrics (11) to itself being a biometric identifier (10). It is not exactly clear what this implies. But in light of the interpretation of the former statement (11), it is reasonable to assume that the photograph is considered to be input in a biometric system—that is, a system that is capable of extracting or generating biometric data from the photograph. Such a clarification also seems to be confirmed in the next statement in this sequence: "In appropriate circumstances, as decided by the issuing State, a facial image can be captured from an endorsed photograph, not requiring the person to be physically present" (9). Admittedly, the word "biometrics" is not used here. It seems nevertheless reasonable to perceive the capturing of the "facial image . . . from an endorsed photograph" as part of such a biometric operation. Let's look a little closer at this ninth statement.

Reading the list from the bottom up, this is the first statement in which the notion "facial image" appears. Unlike the reading of the list as a sequence from the first to the last point, "facial image" here is not established as synonymous with photography. It is, on the contrary, introduced as different from photography as such, as something that can be captured *from* photography without being identical to photography. The notion does not appear in the glossary. The closest one gets is a reference to "full frontal (facial) image," but this is detailed as a photographic portrait.⁴⁵ The word "image," however, is explained as "a representation of a biometric as typically captured via a video, camera or scanning device,"⁴⁶ and "biometric" is defined as "a measurable, unique, physical characteristic or personal behavioral trait used to recognize the identity, or verify the claimed identity, of an enrollee."[47] In the context of the passports, the features in question are physical and machine readable. If we then also look up "machine-verifiable biometric feature" in the glossary, we may seem to have gone full circle, as one of three examples is, exactly, facial image: A "machine-verifiable biometric feature" is "a unique physical personal identification feature (e.g., facial image, fingerprint, or iris) stored electronically in the chip of an eMRTD."[48] The facial image is here considered a biometric feature.

What, then, is implied when it is stated that this biometric feature referred to as the facial image "can be captured from an endorsed photograph"? The word "capture" attracts attention. In *Doc 9303*, "capture" is defined as "the method of taking a biometric sample from the end user."[49] This is normally considered the first step in a facial recognition process. In *Facial Recognition Technology* (2009), Lucas D. Introna and Helen Nissenbaum explain this process in a little more detail:

> The first step in the facial recognition process is the capturing of a face image, also known as the probe image. . . . The effectiveness of the whole system is highly dependent on the quality and characteristics of the captured face image. The process begins with face detection and extraction from the larger image, which generally contains a background and often more complex patterns and even other faces. The system will, to the extent possible, "normalize" (or standardize) the probe image so that it is in the same format (size, rotation, etc.) as the images in the database. The normalized face image is then passed to the recognition software. This normally involves a number of steps such as extracting the features to create a biometric "template" or mathematical representation to be compared to those in the reference database (often referred to as the gallery).[50]

In the ninth statement, the claim is that a biometric sample can be captured from a photograph rather than from a person being physically present. Here, then, it is not a matter of verifying identity (as in the eleventh statement) or of checking identity against a watch list (as in the tenth statement). Without the person of interest being present, this must instead refer to the enrollment of biometric data on a watch list. Instead of capturing the image of a person present, the process involves, in a similar fashion, face detection and extraction from an already existing photograph instead of from an image created at a checkpoint using video or other devices. As Introna and Nissenbaum explain it, the system will then standardize the captured facial image to adjust it in size and rotate it, if necessary, to make it fit the format of the images in the database and make it ready for use in facial recognition software.

Reading the list of statements from the bottom up, the facial image that first appears in this ninth statement is a notion associated not with photography but with *biometry*. Hence, when the next six statements (8–3) then use the term "facial image," it is reasonable to believe that this term still indicates a probe image or biometric sample rather than a photograph.[51] In the next two statements (7–6), it seems clear that the capture of facial images is considered the first step in a biometric operation. It is said that a system for capturing biometric features can be quickly and easily launched (7) and that it would not be expensive to implement such a biometric system for passport issuers (6). The same applies to the next two statements, then, with some consequences we need to examine more carefully.

The fifth statement reports that "the capture of a facial image is nonintrusive. The end user does not have to touch or interact with a physical device for a substantial timeframe to be enrolled." When we read the list from the bottom up, this statement appears not as an appeal to our familiarity with photographs but as an assertion about the nonintrusiveness of FRTs vis-à-vis other biometric modalities. In comparative studies of biometrics, this is in fact often stated without further specifications.[52] If the second part of the statement is supposed to explain the first, however, the nonintrusiveness of FRT seems to be a matter of efficiency and probably also of hygiene and the absence of discomfort.[53] Such an interpretation will not conflict with the aforementioned studies. The nonintrusiveness of this technique could then be read as referring to the contactless image acquisition, which may contribute to higher public acceptance for this modality compared to other candidates.[54]

It is questionable, however, whether this high degree of acceptance can be characterized as informed consent. In their thorough and balanced study of FRTs, Introna and Nissenbaum highlight several advantages of FRT over other biometrics such as fingerprinting, hand geometry, or the iris scan, "as [FRT] imposes fewer demands on subjects and may be conducted at a distance without their knowledge or consent."[55] When applying for a passport, one must expect the public to be "aware of the capture of a facial image and its use for identity verification purposes," as upheld in the fourth statement. This happens routinely as part of the passport application process, as emphasized in the third statement. But do these moral and political considerations, as discussed by Introna and Nissenbaum,⁵⁶ cease by enrollment? If people are not informed that the contactless chip in the passport has an antenna that allows it to be read at a distance of ten centimeters⁵⁷ and compared to a biometric sample (of their facial traits) captured at a gate without their awareness, shouldn't this be considered a political problem? And could it be that this lack of awareness about being controlled as well as about how this control is realized explains some of the relative absence of political opposition to FRT in passports?

Finally, in this reading against the grain of statements supporting the introduction of FRTs in passport inspection systems, we are confronted in the second statement not only with the notion of the facial image as synonymous with photography but also with the sliding transition from photography's social and cultural acceptance internationally to the claim in the very first statement that a facial photograph does not reveal information that the person does not routinely disclose to the general public. There are many studies of these couplings of photography, identification, identity, and face.⁵⁸ Before I wrap this up, let me suggest how these cultural connections are used in the introduction of biometric passports.

By Way of Conclusion: Face as Method

In her book *Our Biometric Future: Facial Recognition Technology and the Culture of Surveillance* (2011), Kelly A. Gates sets out to examine FRT as "unique relative to other forms of biometric identification because the content of the medium is the image of the human face."⁵⁹ She states,

> The technology's use of the face as an object of identification invests it with certain cultural and ideological capital. Facial recognition technology combines an image of high-tech identification with a set of enduring cultural assumptions about the meaning of the face, its unique connection to individuality and identity (in its multiple, conflicting senses), and its distinctive place in human interaction and communication. It is for these reasons, as much as its practical advantages, that facial recognition technology has received special attention.⁶⁰

I find Gates's analysis rich and convincing. But it underestimates two very important factors if we are to understand the introduction of facial biometrics in international passports: an unfounded confidence in the technical accuracy of FRT and the role of photography.

I have argued that the presentation of statements in *Doc 9303* supporting FRT in passport inspection systems reveals that facial biometrics represents a practical, economic, and politically easy way to introduce automated biometrics in passports. Two strategies make this possible. The first is the neglect of the question of technical accuracy when listing the support for a particular biometric modality (FRT), despite the fact that in the same policy document, technical reliability is considered crucial. The absence of comparison with other biometric modalities also helps to make this less visible. The second strategy relates to the cultural history of photography as intimately linked to the histories of individual identity and self-presentation and to the history of identification. By presenting FRT in passports as though it was nothing new compared to photographic pictures, *Doc 9303* facilitates transferring that familiarity with photography to FRT. Reading the list of statements supporting FRT against the grain, however, makes it clear that identification and the verification of identity in passport inspection systems are radically changed when the traveler's appearance is no longer visually compared with a photograph on display in a passport but instead treated as input in a complex system of facial detection and extraction from the larger image, normalization, and a number of steps to enable comparisons with other images. Without going into the complexity of this system, it seems reasonable to argue that most people are not familiar with these technologies. Let me make it clear that I'm not an opponent of biometry per se. But I am a strong supporter of democracy, public debate, and informed consent. As Kristrún Gunnarsdóttir and Kjetil Rommetveit point out, there is no guarantee that public engagement would significantly impact these decisions.[61] Nevertheless, there is conspicuously little debate on the subject, and I have argued that one of the less-explored causes of this may be the choice of the face for biometric input. But why choose facial biometrics if they do not meet the technical requirements?

In this chapter, I have previously announced that the ICAO vision for the application of biometrics technology encompasses some additional considerations that we will return to later. They are about the "specification of agreed supplementary biometric technologies,"[62] that is, fingerprinting and iris scan. There is nothing strange about these specifications, but it seems noteworthy at the end of this analysis to suggest that these additional considerations have great potential. The trend is today, across sectors, that the limitations of monomodal biometric systems are forcing the search for more options. Consequently, the multimodal biometric systems have evolved, as they are often considered to offer better accuracy than are monomodal systems.[63] As underlined by Anil K. Jain, Arun Ross, and Salil Prabhakar, "while small- to medium-scale commercial applications (e.g., a few hundred users) may still use single biometric identification, the only obvious solution for building a highly accurate identification system for large-scale applications appears to be *multimodal biometric* systems."[64] The development of a multimodal system is all the more a matter of course when, as is the case here, the initial solution for a large-scale application has not provided a credible result. It may be worth noting that the ICAO is a consensus-based governance model; as Salter has pointed out, the ICAO is "goal oriented rather than prescriptive."[65] It comes as no surprise then that many countries have been introducing several biometric modalities in passports. Fingerprints, normally not visible on the passport's ID page but included on the chip, are the most common. This has involved even less debate than has the actual introduction of biometric passports. It may therefore be tempting to suggest that the role of the face in biometric passport policy has been to open the door to biometrics in passport inspection systems. The high level of social and cultural acceptance of automated facial recognition made it possible to use FRT as leverage, as a political tool for introducing multimodal biometrics in passports.

Notes

1. International Civil Aviation Organization, *Doc 9303: Machine Readable Travel Documents, Part 1 Machine Readable Passports: Volume 1 Passports with Machine Readable Data Stored in Optical Character Recognition Format*, 6th ed. (Montreal: International Civil Aviation Organization, 2006), I-2.

2. "Over 60+ Countries Now Issuing EPassports," FindBiometrics, accessed 4 June 2018, https://findbiometrics.com/over-60-countries-now-issuing-epassports-2/.

3. Gemalto reported that 120 countries were issuing ePassports as of mid-June 2017, and this information is presented in an updated article from 22 May 2018. See "The Electronic Passport in 2018 and Beyond," Gemalto, accessed 4 June 2018, https://www.gemalto.com/govt/travel/electronic-passport-trends. Most of the Central American countries as well as some African and South Asian countries have not yet implemented biometric passports. The fact that many of these are poor countries may indicate that the introduction of biometric passports entails high costs and infrastructural development.

4. International Civil Aviation Organization, *Doc 9303: Machine Readable Travel Documents, Part 1 Introduction*, 7th ed. (Montreal: International Civil Aviation Organization, 2015), 1.

5. Jeffrey M. Stanton, "ICAO and the Biometric RFID Passport," in *Playing the Identity Card: Surveillance, Security and Identification in Global Perspective*, ed. Colin J. Bennet and David Lyon (London: Routledge, 2008), 262.

6. Mark B. Salter and Can E. Mutlu, "Psychoanalytic Theory and Border Security," *European Journal of Social Theory* 15, no. 2 (2011): 189.

7. Kristrún Gunnarsdóttir and Kjetil Rommetveit, "The Biometric Imaginary: (Dis)trust in Policy Vacuum," *Public Understanding of Science* 26, no. 2 (2017): 196.

8. Ibid., 198–99. See also Jonathan P. Aus, "EU Governance in an Area of Freedom, Security and Justice: Logics of Decision-Making in the Justice and Home Affairs Council," ARENA Working Paper 2007.15, ARENA Centre for European Studies, Faculty of Social Sciences, University of Oslo, October 2007, https://www.sv.uio.no/arena/english/research/publications/arena-working-papers/2001-2010/2007/wp07_15.pdf; and Salter and Mutlu, "Psychoanalytic Theory and Border Security," 180.

9. Stanton, "ICAO and the Biometric RFID Passport," 259–60. See also Gunnarsdóttir and Rommetveit, "Biometric Imaginary," 196.

10. See, for instance, Aus, "EU Governance," 30.

11. Ibid.

12. Gunnarsdóttir and Rommetveit, "Biometric Imaginary," 196. See also Aus, "EU Governance," 30.

13. See, for example, Dyala R. Ibrahim, Abdelfatah A. Tamimi, and Ayman M. Abdalla, "Performance Analysis of Biometric Recognition Modalities" (paper presentation, 8th International Conference on Information Technology [ICIT], May 2017, 983), https://doi.org/10.1109/ICITECH.2017.8079977; Gursimapreet Kaur and Chander Kant Verma, "Comparative Analysis of Biometric Modalities," *International Journal of Advanced Research in Computer Science and Software Engineering* 4, no. 4 (2014): 611; Emilio Mordini and Andrew P. Rebera, "The Biometric Fetish," in *Identification and Registration Practices in Transnational Perspective: People, Papers and Practices*, eds. Ilsen About, James Brown, and Gayle Lonergan (Basingstoke, UK: Palgrave Macmillan, 2013), 100–101; Nancy Yue Liu, *Bio-Privacy: Privacy Regulations and the Challenge of Biometrics* (Abingdon, UK: Routledge, 2013), 41, 81; Kelly A. Gates, *Our Biometric Future: Facial Recognition Technology and the Culture of Surveillance* (New York: New York University Press, 2011), 17; and Julian Ashbourn, *Guide to Biometrics for Large-Scale Systems: Technological, Operational, and User-Related Factors* (London: Springer, 2011), 22.

14. Ibrahim, Tamimi, and Abdalla, "Performance Analysis," 983; Kaur and Verma, "Comparative Analysis," 606, 610; Anil K. Jain and Ajay Kumar, "Biometric Recognition: An Overview," in *Second Generation Biometrics: The Ethical, Legal and Social Context*, ed. Emilio Mordini and Dimitros Tzovaras (Dordrecht, NL: Springer, 2012), 50; Anil K. Jain, Arun Ross, and Salil Prabhakar, "An Introduction to Biometric Recognition," *IEEE Transactions on Circuits and Systems for Video Technology* 14, no. 1 (January 2004): 11.

15. The ICAO is a specialized agency of the United Nations charged with developing the principles and techniques of international air navigation. The agency has the technical mandate to regulate the worldwide air travel infrastructure. These regulations are binding on the member states, of which there are currently 192. The official list of ICAO Contracting States is at "Member States," ICAO, accessed 2 June 2018, https://www.icao.int/about-icao/Pages/member-states.aspx. If we count 195 countries in the world today, the ICAO covers most of the globe. "How Many Countries Are There in the World?," Worldometers, accessed 2 June 2018, http://www.worldometers.info/geography/how-many-countries-are-there-in-the-world/. Instead of studying regulations in the European Union or United States, for example, or investigating national implementations of international regulations, I will emphasize the importance of

examining the ICAO-endorsed international standards for biometric passports.

16. Paul Ricoeur, "The Model of the Text: Meaningful Action Considered as a Text," *New Literary History* 5, no. 1 (Autumn 1973): 95. Without changing Ricoeur's points substantively, I have slightly adjusted the quote from "the psychology of its author" to "the intention of its author," which seems more appropriate for policy documents.

17. See ibid., 103.

18. International Civil Aviation Organization, *Doc 9303: Machine Readable Travel Documents, Part 9 Deployment of Biometric Identification and Electronic Storage of Data in eMRTDs*, 7th ed. (Montreal: International Civil Aviation Organization, 2015), 7. I am quoting the seventh and most recent edition of *Doc 9303*, published in late 2015. This version of the document is reorganized and divided into twelve parts. The regulation of biometric information is specified in Part 9. These were not substantively modified from the previous edition. The acronym eMRTD is an abbreviation of "electronic Machine-Readable Travel Document."

19. Ibid.

20. Ibid., 4.

21. International Civil Aviation Organization, *Introduction*, 16.

22. Lucas D. Introna and Helen Nissenbaum, *Facial Recognition Technology: A Survey of Policy and Implementation Issues* (New York: Center for Catastrophe Preparedness and Response, New York University, 2009), 13.

23. International Civil Aviation Organization, *Deployment of Biometric Identification*, 4.

24. Introna and Nissenbaum, *Facial Recognition Technology*, 12.

25. See ibid., 18.

26. International Civil Aviation Organization, *Deployment of Biometric Identification*, 4. This vision encompasses some additional considerations that we will return to later.

27. Ibid.

28. Gates, *Our Biometric Future*, 17.

29. Ashbourn, *Guide to Biometrics*, 22.

30. Mordini and Rebera, "Biometric Fetish," 100–101.

31. Jain, Ross, and Prabhakar, "Introduction to Biometric Recognition," 9.

32. Both goals are underlined in *Doc 9303*, for example. See International Civil Aviation Organization, *Introduction*, 2.

33. International Civil Aviation Organization, *Deployment of Biometric Identification*, 7–8. This list was first presented in 2006. In 2015, bullet points replaced the numbering. Except for a few minor details in the third and eighth statements, the wordings are reproduced and the list is presented in the same order. To make it easier to refer to the various points in the analysis, I find it appropriate to number the eleven statements and reference them parenthetically within the chapter.

34. International Civil Aviation Organization, *Doc 9303*, 7th ed., 7.

35. FRT refers to a biometric system for recognition of facial features. The biometric information is not always photographically generated. Admittedly, cameras or scanning devices are involved in the production and verification of both the mandatory (facial biometrics) and two optional (fingerprint biometric technology and iris biometrics) biometric modalities in biometric passports. Yet the recognition technology as such is a complex system that involves much more than a photographic image. As Allen Sekula already so aptly said about Alphonse Bertillon's anthropometric system from the late 1880s, the bertillonage, "the camera is integrated into a larger ensemble: a bureaucratic-clerical-statistical system of 'intelligence.' . . . The central artifact of this system is not the camera but the filing cabinet." "The Body and the Archive," *October* 39 (Winter 1986): 16. Needless to say, the FRT involved in today's passport inspection systems is far more complex than that of the bertillonage. For further discussion, see my article "The Archival Promise of the Biometric Passport," in *Memory in Motion: Archives, Technology and the Social*, ed. Ina Blom, Eivind Røssaak, and Trond Lundemo (Amsterdam: Amsterdam University Press, 2016), 257–84.

36. Jane Caplan, "'This or That Particular Person': Protocols of Identification in Nineteenth-Century Europe," in *Documenting Individual Identity: The Development of State Practices in the Modern World*, ed. Jane Caplan and John Torpey (Princeton: Princeton University Press, 2001), 50. See also Gates, *Our Biometric Future*, 15.

37. Ricoeur, "Model of the Text," 107.

38. International Civil Aviation Organization, *Introduction*, 20.

39. Salter and Mutlu, "Psychoanalytic Theory and Border Security," 190.

40. Robin Cooper Feldman, "Considerations on the Emerging Implementation of Biometric Technology," *25 Hastings Communication and Entertainment Law Journal* 653 (2003): 666, https://repository.uchastings.edu/faculty_scholarship/160.

41. Ashbourn, *Guide to Biometrics*, 114.

42. International Civil Aviation Organization, *Introduction*, 7.

43. Ibid., 8.

44. For a discussion of the relationship between visual photographs and the photographically generated information stored on the chip in today's passport, see my article "Photographic Passport Biometry," *PUBLIC: Art/Culture/Ideas* 60 (April 21, 2020): 45–53.

45. "A portrait of the holder of the MRTD produced in accordance with the specifications established in Doc 9303." International Civil Aviation Organization, *Introduction*, 12.

46. International Civil Aviation Organization, *Introduction*, 13.

47. Ibid., 7.

48. Ibid., 15.

49. Ibid., 8.

50. Introna and Nissenbaum, *Facial Recognition Technology*, 11.

51. The "facial images" mentioned in the eighth statement probably appear as digital photographs, given the reference to legacy databases. These digital files seem to be referred to here as possible sources for biometric comparison with new images—that is, probe images.

52. Ibrahim, Tamimi and Abdalla, "Performance Analysis," 981; Jain, Ross, and Prabhakar, "Introduction to Biometric Recognition," 9; Kanupriya Sharma and Aseem Kumar, "Review of Study on Comparative Analysis of Biometric Authentication Security for M-commerce," *International Journal of Advanced Research in Computer Science and Software Engineering* 4, no. 1 (January 2014): 1164; Jawed Akhtar Unar, Woo Chaw Seng, and Almas Abbasi, "A Review of Biometric Technology Along with Trends and Prospects," *Pattern Recognition* 47 (2014): 2677.

53. See Chris Riley et al. for people's concern about hygiene in relation to biometrics. The question of discomfort has been particularly important in the assessment of iris scanning. "Culture & Biometrics: Regional Differences in the Perception of Biometric Authentication Technologies," *AI and Society* 24, no. 3 (October 2009): 296.

54. See, for instance, Unar, Seng, and Abbasi, "Review of Biometric Technology," 2677.

55. Introna and Nissenbaum, *Facial Recognition Technology*, 21.

56. See particularly ibid., 5.

57. See International Civil Aviation Organization, *Deployment of Biometric Identification*, 11.

58. See, for instance, Gates, *Our Biometric Future*; Jonathan Finn, *Capturing the Criminal Image: From Mug Shot to Surveillance Society* (Minneapolis: University of Minnesota Press, 2009); and Hans Belting, *Face and Mask: A Double History*, trans. Thomas S. Hansen and Abby J. Hansen (Princeton: Princeton University Press, 2017).

59. Gates, *Our Biometric Future*, 17.

60. Ibid., 18.

61. Gunnarsdóttir and Rommetveit, "Biometric Imaginary," 197.

62. International Civil Aviation Organization, *Deployment of Biometric Identification*, 4.

63. See, for instance, Unar, Seng, and Abbasi, "Review of Biometric Technology," 2684.

64. Jain, Ross, and Prabhakar, "Introduction to Biometric Recognition," 13 (emphasis in original).

65. Mark B. Salter, "The Global Airport: Managing Space, Speed, and Security," in *Politics at the Airport*, ed. Mark B. Salter (Minneapolis: University of Minnesota Press, 2008), 14. Some scholars have questioned the ICAO's authority to set and enforce international policy (see Stanton, "ICAO and the Biometric RFID Passport"). Given that it is an international organization of sovereign states, the ICAO indeed has some limitations "with respect to its ability to impose regulations, procedures and standard practices on matters occurring within the bounds of the sovereign states that comprise its membership" (Stanton, "ICAO and the Biometric RFID Passport," 257; see also Salter, "Global Airport," 16). Some have also questioned the legitimacy of the organization on the basis of what is believed to be the disproportionately high influence of the United States on its policy. See Stanton, "ICAO and the Biometric RFID Passport"; Salter and Mutlu, "Psychoanalytic Theory and Border Security," 190; and Jane Caplan and Edward Higgs, "Afterword: The Future of Identification's Past: Reflections on the Development of Historical Identification Studies," in *Identification and Registration Practices in Transnational Perspective: People, Papers and Practices*, ed. Ilsen About, James Brown, and Gayle Lonergan (Basingstoke, UK: Palgrave Macmillan, 2013), 304. Despite the potential tensions between the ICAO and its 192 member states, the ICAO has had during its more than seventy-year history the regulatory authority and responsibility for reaching consensus on international civil aviation standards, recommended practices, and policies "in support of a safe, efficient, secure, economically sustainable and environmentally responsible civil aviation sector." (See the ICAO's official site: "About ICAO," ICAO, accessed 25 June 2018, https://www.icao.int/about-icao/Pages/default.aspx.)

References

Ashbourn, Julian. *Guide to Biometrics for Large-Scale Systems: Technological, Operational, and User-Related Factors*. London: Springer, 2011.

Aus, Jonathan P. "EU Governance in an Area of Freedom, Security and Justice: Logics of Decision-Making in the Justice and Home Affairs Council." ARENA Working Paper 2007.15, ARENA Centre for European Studies, Faculty of Social Sciences, University of Oslo, October 2007. https://www.sv.uio.no/arena/english/research/publications/arena-working-papers/2001-2010/2007/wp07_15.pdf.

Belting, Hans. *Face and Mask: A Double History*. Translated by Thomas S. Hansen and Abby J. Hansen. Princeton: Princeton University Press, 2017.

Caplan, Jane. "'This or That Particular Person': Protocols of Identification in Nineteenth-Century Europe." In *Documenting Individual Identity: The Development of State Practices in the Modern World*, edited by Jane Caplan and John Torpey, 49–66. Princeton: Princeton University Press, 2001.

Caplan, Jane, and Edward Higgs. "Afterword: The Future of Identification's Past: Reflections on the Development of Historical Identification Studies." In *Identification and Registration Practices in Transnational Perspective: People, Papers and Practices*, edited by Ilsen About, James Brown, and Gayle Lonergan, 302–8. Basingstoke, UK: Palgrave Macmillan, 2013.

Feldman, Robin Cooper. "Considerations on the Emerging Implementation of Biometric Technology." *25 Hastings Communication and Entertainment Law Journal* 653 (2003): 653–82. https://repository.uchastings.edu/faculty_scholarship/160.

FindBiometrics. "Over 60+ Countries Now Issuing EPassports." Accessed 4 June 2018. https://findbiometrics.com/over-60-countries-now-issuing-epassports-2/.

Finn, Jonathan. *Capturing the Criminal Image: From Mug Shot to Surveillance Society*. Minneapolis: University of Minnesota Press, 2009.

Gates, Kelly A. *Our Biometric Future: Facial Recognition Technology and the Culture of Surveillance*. New York: New York University Press, 2011.

Gemalto. "The Electronic Passport in 2018 and Beyond." Accessed 4 June 2018. https://www.gemalto.com/govt/travel/electronic-passport-trends.

Gunnarsdóttir, Kristrún, and Kjetil Rommetveit. "The Biometric Imaginary: (Dis)trust in Policy Vacuum." *Public Understanding of Science*, 26, no. 2 (2017): 195–211.

Hausken, Liv. "The Archival Promise of the Biometric Passport." In *Memory in Motion: Archives, Technology and the Social*, edited by Ina Blom, Eivind Røssaak, and Trond Lundemo, 257–84. Amsterdam: Amsterdam University Press, 2016.

———. "Photographic Passport Biometry." *PUBLIC: Art/Culture/Ideas* 60 (April 21, 2020): 45–53.

Ibrahim, Dyala R., Abdelfatah A. Tamimi, and Ayman M. Abdalla. "Performance Analysis of Biometric Recognition Modalities." Paper presented at the *8th International Conference on Information Technology (ICIT)*, Amman, Jordan, May 2017. https://doi.org/10.1109/ICITECH.2017.8079977.

International Civil Aviation Organization. "About ICAO." ICAO. Accessed 25 June 2018. https://www.icao.int/about-icao/Pages/default.aspx.

———. *Doc 9303: Machine Readable Travel Documents, Part 1 Introduction*. 7th ed. Montreal: International Civil Aviation Organization, 2015.

———. *Doc 9303: Machine Readable Travel Documents, Part 1 Machine Readable Passports: Volume 1 Passports with Machine Readable Data Stored in Optical Character Recognition Format*. 6th ed. Montreal: International Civil Aviation Organization, 2006.

———. *Doc 9303: Machine Readable Travel Documents, Part 1 Machine Readable Passports: Volume 2 Specifications for Electronically Enabled Passports with Biometric Identification Capability*. 6th ed. Montreal: International Civil Aviation Organization, 2006.

———. *Doc 9303: Machine Readable Travel Documents, Part 9 Deployment of Biometric Identification and Electronic Storage of Data in eMRTDs*. 7th ed. Montreal: International Civil Aviation Organization, 2015.

———. "Member States." ICAO. Accessed 2 June 2018. https://www.icao.int/about-icao/Pages/member-states.aspx.

Introna, Lucas D., and Helen Nissenbaum. *Facial Recognition Technology: A Survey of Policy and Implementation Issues*. New York: Center for Catastrophe Preparedness and Response, New York University, 2009.

Jain, Anil K., and Ajay Kumar. "Biometric Recognition: An Overview." In *Second Generation Biometrics: The Ethical, Legal and Social Context*, edited by Emilio Mordini and Dimitros Tzovaras, 49–79. Dordrecht, NL: Springer, 2012.

Jain, Anil K., Arun Ross, and Salil Prabhakar. "An Introduction to Biometric Recognition." *IEEE Transactions on Circuits and Systems for Video Technology* 14, no. 1 (January 2004): 4–20.

Kaur, Gursimapreet, and Chander Kant Verma. "Comparative Analysis of Biometric Modalities." *International Journal of Advanced Research in Computer Science and Software Engineering* 4, no. 4 (2014): 603–13.

Liu, Nancy Yue. *Bio-Privacy: Privacy Regulations and the Challenge of Biometrics*. Abingdon, UK: Routledge, 2013.

Mordini, Emilio, and Andrew P. Rebera. "The Biometric Fetish." In *Identification and Registration Practices in Transnational Perspective: People, Papers and Practices*, edited by Ilsen About, James Brown, and Gayle Lonergan, 98–110. Basingstoke, UK: Palgrave Macmillan, 2013.

Ricoeur, Paul. "The Model of the Text: Meaningful Action Considered as a Text." *New Literary History* 5, no. 1 (Autumn 1973): 91–117.

Riley, Chris, Kathy Buckner, Graham Johnson, and David Benyon. "Culture & Biometrics: Regional Differences in the Perception of Biometric Authentication Technologies." *AI and Society* 24, no. 3 (October 2009): 295–306.

Salter, Mark B. "The Global Airport: Managing Space, Speed, and Security." In *Politics at the Airport*, edited by Mark B. Salter, 1–28. Minneapolis: University of Minnesota Press, 2008.

Salter, Mark B., and Can E. Mutlu. "Psychoanalytic Theory and Border Security." *European Journal of Social Theory* 15, no. 2 (2011): 179–95.

Sekula, Allen. "The Body and the Archive." *October* 39 (Winter 1986).

Sharma, Kanupriya, and Aseem Kumar. "Review of Study on Comparative Analysis of Biometric Authentication Security for M-commerce." *International Journal of Advanced Research in Computer Science and Software Engineering* 4, no. 1 (January 2014): 1161–65.

Stanton, Jeffrey M. "ICAO and the Biometric RFID Passport." In *Playing the Identity Card: Surveillance, Security and Identification in Global Perspective*, edited by Colin J. Bennet and David Lyon, 253–67. London: Routledge, 2008.

Unar, Jawed Akhtar, Woo Chaw Seng, and Almas Abbasi. "A Review of Biometric Technology Along with Trends and Prospects." *Pattern Recognition* 47 (2014): 2673–88.

Worldometers. "How Many Countries Are There in the World?" Accessed 2 June 2018. http://www.worldometers.info/geography/how-many-countries-are-there-in-the-world/.

9.
The Bourne Identification

Rex Ferguson

These people know who I am.
—*The Bourne Identity*

The value that modern Western society has placed on individual freedom has often made it appear a primary "good." Within a longer history of a modernity in which the bourgeois worldview rested on the autonomy of the sovereign individual, more recent movements and stances have coalesced around the rights of individuals to have their own independently construed life choices respected and guaranteed. That these should only be infringed when a necessity such as the preservation of life is in question has also been the standard rhetoric that has animated armed conflicts and, more recently, antiterrorism legislation. Within this context, the techniques employed to definitively identify human beings that have taken on an ever-greater prominence in the developed world since around the mid-nineteenth century are often railed against as essentially limiting acts that do violence to the autonomous individual's capacity to self-create.[1] Literary history is full of examples in which the individual, in running up against his or her own identification, is prompted to feel, often in quite stark terms, the essential reductiveness of the process.[2] Taking this position as a starting point and beginning with the United Kingdom as an example before moving on to a more expansive plain, this chapter suggests that the resistance to identification tells only half the story. For running alongside the desire for freedom from the various institutions and organizations that "name"—and a history that prides itself on the protection of individual liberty—contemporary society in the United Kingdom and other developed nation-states is in many ways dominated by systems of identification and surveillance. In seeking to illustrate a potentially universal explanation for this, this chapter will turn to an analysis of the Jason Bourne series of films, released between 2002 and 2016. What these films demonstrate, in a dramatization that is exaggerated to the point of fantasy, is that the submission to systems of identification evident in contemporary society is not the result of a rational calculation in which the cost of a reduction in freedom is measured against national security. Rather, the character of Jason Bourne exemplifies the messy conditions that apply when we do not just accept less freedom for more security but want to be both anonymous and identifiable at the same time—when the desire to be free is met by the desire to be named.

Identification and Its Discontents

The mythology that surrounds the United Kingdom's valorization of individual liberty (and its undermining) is, in terms of recent history, wrapped up in the popular myth of World War II.[3] The war is collectively remembered as a fight for freedom in which military personnel and those at home worked together freely to defeat tyranny. Yet what is normally elided in this myth is the litany of quite draconian emergency legislation that was passed during the first few weeks of the conflict.[4] Among these new laws was the institution of a National Register and the issuing to all citizens of National Registration Identity Cards that had to be kept on the person at all times. Despite being related to the "emergency" of the war, this legislation remained in force for some six years following the armistice. For many, this made perfect sense. An editorial in *The People* on 17 June 1951 read that "if this National Registration system were rigidly enforced, we could round up half our deserters, keep an eye on more than half our spies and 'fellow travellers,' reduce the shocking rate of bigamy in Britain—and cause no trouble at all to ordinary folk."[5] The logic of this statement is read as that only wrongdoers will suffer from blanket identification, while law-abiding citizens have nothing to fear. Others did not agree. On 7 December 1950, Mr. Clarence Henry Willcock (a member of the Liberal Party) pushed the issue when he refused to produce his card for Police Constable Harold Muckle. Willcock was charged and found guilty of being in breach of section 6 (4) of the National Registration Act of 1939, though the court withheld punishment.[6] When Willcock appealed, the case went before the High Court of Justice, King's Bench Division. The higher court upheld Willcock's conviction but also endorsed the decision not to punish, and the final judgment was highly critical of the government's position, arguing that the National Registration Act "was passed for security purposes; it was never passed for the purposes for which it is now apparently sought to be used." In the court's view, allowing the police to demand identity cards from "all and sundry" even in the most innocuous situations was "wholly unreasonable" and "tends to turn law-abiding subjects into lawbreakers, which is a most undesirable state of affairs."[7] Much of the press agreed, and the pressure that ensued in the aftermath of the Willcock case ensured that the requirement to carry identity cards was soon abandoned.[8]

The debate generated by World War II identity cards is, in many ways, entirely rational. Cards are useful, clearly, but they are also, literally and according to the legislation, unnecessary once the nation is no longer at war. Yet rather than restricting their judgment to the particulars of the case or even to the technical question of defining precisely when the "state of emergency" was deemed to have ended, the High Court evidently saw a matter of principle at stake. In its words, to be asked to prove one's identity is "wholly unreasonable," while to be branded a "lawbreaker" simply by one's anonymity is "undesirable." The government's need to know all its citizens—and what it might do with that knowledge and the right for it to be rendered from the individual—fed into the wider myth of the war itself and, though necessary during wartime, was now eerily reminiscent of the totalitarian enemy that had been fought.

This history has played a part in the United Kingdom's resistance to the imposition of a national identity card to this day—something that has set it apart from many of its European neighbors. There have been attempts to reimagine the system, though. Just over half a century after their wartime iteration had been scrapped, the Labour government, led by Tony Blair, launched a new plan for national identity cards, the legislation for which was passed with the Identity Cards Act of 2006. The issuing of cards and the setting of an accompanying register were intended to achieve several aims including reducing identity theft, disrupting terrorist activity, and tackling benefit fraud. Blair also wrote in a Cabinet Office report of 2005 that "government will create an holistic approach to identity management, based on a suite of identity management solutions that enable the public and private sectors to manage risk and provide cost-effective services trusted by customers and stakeholders."[9] The adoption of a commercial language of management and service prompted a critique based on the same rhetoric used to question the financial implications and effectiveness of the plans. Despite assurances to the contrary, worry that the demand to present one's card would inevitably become a part of police practices of stop-and-search and thus disproportionately affect young ethnic minorities also led to important objections. But in making plans for the convergence of identity management "toward biometric identity cards and the National Identity Register," the Labour government's plans came up against what appeared to be historically entrenched instincts.[10] First, identity cards were intrinsically associated with a state of emergency. Second, in utilizing the body as biometric data with which to identify the individual, identity cards were associated with the identification of criminals. Edward Higgs writes that "when identifying themselves as juridical persons, who are able to claim the right to acquire and alienate property, individuals have historically used possessions they could produce, acts they could perform, or signs they could utter. . . . On the other hand, criminals, and other deviants, have historically been identified through the body."[11] The identity card, while still retained as a possession, was grounded in the inalienable signs of a body that would give itself away and was thus powerfully resonant of criminalization. As such, the Labour government's plan faltered, and the act was repealed in 2010.

What makes this latter manifestation of identity cards so significant is that the apparent assertion of individual autonomy—the refusal to be "known" by the state—takes place in a society in which citizens are identified and monitored in strikingly intrusive ways. For example, David Barnard-Wills reports that the "UK is regularly named the most surveilled country in the world, with the Square Mile of the city of London the most surveilled public space in the world."[12] The United Kingdom also has, per capita, the largest DNA database in the world.[13] To envisage the United Kingdom as a bastion of liberty, as a nation with a proud history of defending individual freedoms, is thus to ignore the evident acquiescence to the structures of surveillant identification that have seeped into contemporary society. The establishment of protocols around the use of DNA and the populating of the national DNA database is particularly instructive on this front, partly because new issues are continually evolving. As recently as July 2018, for instance, the Home Office admitted to the legally "dubious" practice used in taking DNA swabs from those seeking asylum in the United Kingdom.[14] Between 2003 and 2008, DNA profiles were also generated for all those arrested by the police, regardless of whether they were ultimately charged with a crime—a practice that came to an end only following the European Court of Human Rights ruling in *S & Marper v. United Kingdom* (2008) and that still pertains to serious crimes such as rape or murder. DNA profiles, though limited in their scope and restricted to noncoding regions of the genome, have also allowed for familial searching in which a sample taken from a crime scene is partially matched to a profile held on the database. When that match is close enough, siblings of the individual who generated the profile, even though they had never given a DNA sample to the authorities before, can be rendered as suspects. The United Kingdom has been a pioneer of such searching, with notable cases having been solved due to its application.[15] Sir Alec Jeffreys, who pioneered the technique of DNA profiling in 1984, has recently expressed concern about the lack of legislation in this area, claiming that profiling has raised "ethical issues that still haven't been fully resolved." Summarizing the libertarian position with some clarity, Jeffreys takes "the very simple view that my genome is my personal property and is not up for grabs nor is available for free and open viewing by any organisation."[16]

Of course, even in a context in which the United Kingdom's DNA database is both the largest per capita in the world and includes the profiles of those never charged with a crime, it still represents a small percentage of the entire population (around 5 percent).[17] The resonance that it has is amplified, however, by its exemplary status within a portfolio of techniques that have made biometric information an intrinsic part of contemporary surveillance culture in the majority of developed nation-states. Several police forces around the world now use portable fingerprint scanners that can run prints against a networked database, while iris scanning has been used by certain border control authorities, employers, and banks for some time.[18] Facial recognition technology has also become increasingly effective and is used in a variety of contexts.[19] For instance, William G. Staples writes of an area of East London in which "not only does the system videotape people walking down the street or coming out of shops, but, using fractal processing, it is able to isolate, recognize, and identify an individual's facial characteristics and compare them with those stored in a database."[20]

The fact that biometric technologies are being used in this manner complicates Higgs's useful distinction between historical modes of identification. Rather than being neatly separated, employees, customers, and those engaged in lawful transit are identified by their bodies, while many of the tokens of identification that can be freely proffered are now laced with biometric data.[21] Irma van der Ploeg writes that "the coupling of biometrics with

IT unequivocally puts the body center stage," but there is a question about what the body becomes through this combination.[22] With iris scanning, for example, the body is envisaged in a form that is entirely dependent on the technological means through which an identifying pattern is generated. And even when technology such as a portable fingerprint scanner appears to just extend a practice traditionally carried out with ink and paper, it fundamentally does so by translating a physical pattern of lines into the ones and zeros of digital code. For Katja Aas, this translation "radically reduces possibilities for negotiation and therefore also resistance. In DNA profiles, biometric scans, X-ray photographs and drug tests the body is a source of information that is standardized and unambiguous in its meaning."[23]

Aas assumes that resistance is the goal—that the rejection of definitive identification is a right to be fought for—but there is an alternative view. Higgs's work on the history of identification, for instance, is premised on the thesis that practices of identification have been focused on issues of criminality *and* the rights of citizens. He claims that "the right to identification can thus be a right to be recognized."[24] For David Lyon, the issue of recognition, or lack thereof, is similarly key; he argues that "whereas face-to-face contacts have characterized all previous human history, today they no longer predominate. While some remain, of course, most relationships are now mediated by many means.... Bodies are disappearing."[25] Read in this light, the relentless tide of moments that require individuals to identify themselves through the insertion of codes, the validation of usernames, and the swiping of cards responds to a deep-seated need to recognize and be recognized. Paradoxically, then, it is the very mediation that reduces the face-to-face nature of human relations that becomes the means of producing a different kind of familiarity. Mark Poster comments on this when writing that

> previously anonymous actions such as paying for a dinner, borrowing a book from the library, renting a videotape from a rental store, subscribing to a magazine, making a long distance telephone call—all by interacting with perfect strangers—now are wrapped in a clothing of information traces which are gathered and arranged into profiles.... In the credit card payment for dinner, the waiter is a stranger but the computer which receives the information "knows" the customer very well.[26]

That these constant identifications also produce traceable records of movement and activity has amplified the surveillant nature of contemporary experience, rendering citizens "known" through a coordination of the "fragmented, ephemeral, almost inconsequential" aspects of quotidian life.[27] Poster's examples are of actions that not only identify but also record presence as a digital trace capable of retrieval. Every single time an identification takes place, through a debit-card transaction or the swiping of a travel card, for example, a trace is stored of that individual's presence in a particular location at a particular time. Added to which could be all the moments that are not overtly connected to a process of identification but that hold the potential to be reconstructed as such at a later date—visits to websites, walks through urban centers that are monitored by CCTV, or conversations on the telephone, for instance.

After the blurring of identification through body or token, what is noticeable about this contemporary mode of surveillance is the way it extends far beyond state-run activity. Lyon thus writes of "the mundane, ordinary, taken-for-granted world of getting money from a bank machine, making a phone call, applying for sickness benefits, driving a car, using a credit card, receiving junk mail, picking up books from the library, or crossing a border on trips abroad."[28] Within a hypercapitalist and globalized world, identifying information has become a valuable commodity for organizations (commercial and state run) that conceive of individuals as customers who are best served through a tailored, bespoke experience. Indeed, this feature is what prompted Kevin D. Haggerty and Richard V. Ericson's influential formulation that contemporary experience is conducted via a "surveillant assemblage" that is "rhizomatic" in its construction.[29] The assumption of Deleuze and Guattari's terminology here is intended to emphasize the notion that the multiplying means of identifying and surveilling an individual are generative not of a single "grand narrative" but of a proliferation of identities. It is important to recognize, though, how this assemblage is capable of being mobilized in the service of the state at any given moment and that, as the Edward Snowden revelations demonstrated, modern nation-states are deeply invested in the surveillant potential offered by, for example, telecommunications companies.

That such revelations do not appear to have had a radical effect on the behavior of citizens is partially explained by the subtle forms of coercion that characterize contemporary experience. Clare Birchall, for instance, argues that internet activity forces citizens to "'share' data with the veillant state in a way that renders them visible and trackable."[30] But what is equally and perhaps even more strikingly evident is the compliance that most citizens demonstrate in the face of such surveillance. Lyon thus writes of a "reciprocal" arrangement in which "we collude more and more with our own surveillance."[31] Christian Parenti goes further, arguing that "we are not 'being watched' so much as we are voluntarily 'checking in' with authorities. If looked at from this view, the landscape is now littered with registration kiosks—ATMs, automatic ticketing machines, electronic tolls—where we deposit personal information in exchange for services."[32] "Service," in this arrangement, is facilitated through a recording of the past that generates a virtual future of imagined possibility. In other words, the "registration" kiosks that are freely checked into are connected to vast databases of information that self-generate predictive profiles of what each user will need or want in the future.[33] From a commercial angle, this form of profiling allows for targeted marketing. From the position of the state, it represents a vast repository of information that assists in the generation of comprehensive risk assessments once a given individual comes to the attention of the security services.[34] That this leads to a conceptualization of individuality as a collection of elements and traces is certainly not an entirely positive development, not least because of the detrimental effect it may have on the liberty and autonomy of individual citizens. But it is the price one must pay, not for safety or security but to fulfill the pressing desire to be recognized in the contemporary world.

Bourne's Web(b)

The series of films starring Matt Damon titled *The Bourne Identity* (2002), *The Bourne Supremacy* (2004), *The Bourne Ultimatum* (2007), and *Jason Bourne* (2016) begins with an apparently dead body floating in water that a caption announces is the "Mediterranean Sea 60 miles south of Marseilles."[35] Pulled out by Italian fishermen, the unknown male body turns out not to be dead, though he has been shot twice in the back. When he regains consciousness, the central conceit of the series is revealed: suffering from amnesia, this man is as unknown to himself as he is to the viewer.[36] What emerges during the first film in the series is that this man is a highly trained and behaviorally modified CIA operative for a project named Treadstone who lives under the assumed name of Jason Bourne and has spent much of the recent past carrying out covert assassinations around the world. This revelation takes place gradually, as the narrative of the film develops, and forms a neat structural coherence in which Bourne's discoveries are, in the main, made at the same time as those of the audience. This leads inexorably to an identification on the part of the viewer with Bourne, an identification enhanced by a seemingly obvious fact: namely, that Bourne's journey is also a more personal and even universal one of self-discovery in which he returns to his original, "real" identity.[37] The task of "knowing himself" will thus truly be fulfilled only once he regains knowledge of who he was before becoming Bourne—a task that inevitably involves evading and outsmarting the authorities who seek both to identify his movements and to reassert his identification as Bourne.

That the Bourne series is about resisting the capacity held by institutions and organizations to identify and name is made clear from ample evidence in the films themselves. From the very beginning, the individual is thus represented as the intensely vulnerable object of big government (represented by the CIA) and its control of the "surveillant assemblage"—referred to in the films as the "grid." Bourne's first clue about his identity comes through the retrieval of an implant under the skin on his hip; the tiny laser projector he finds there provides the details of a safety deposit box held in a Zurich bank. The penetration of and embedding of information in Bourne's body is then instantly matched by the way his body is embedded, as information, in a network of surveillant devices. Bourne's access to the room in which he views the contents of the safety deposit box is, for instance, gained through placing his hand on an electronic scanner—enacting the kind of moment written about by Poster, in which Bourne is "known" by the device, while he is anonymous to the staff and to himself. His resurrected presence is flagged up to the CIA, and then the man-on-the-run narrative begins, as Bourne's primary task is the evasion of state-monitored surveillance, images of which then dominate each film. Through a series of open-plan control rooms, located either centrally at CIA headquarters in Langley, Virginia, or in Berlin, London, and New York, Bourne's movements are the target of a succession of agents—Alexander Conklin (*Identity*), Pamela Landy (*Supremacy* and *Ultimatum*), Noah Vosen (*Ultimatum*), and Robert Dewey (*Jason Bourne*). In *Identity*, the vulnerability of the individual in the face of such panopticism is rendered through Marie, a young woman whom Bourne pays to drive him to Paris and who later becomes his partner. Identifying her from CCTV footage in Zurich, Conklin orders that her past be traced, and almost instantly a scene follows in which a low-ranking agent outlines her family history and movements. Significantly, while Marie's record is characterized as "chaotic at best," she still "pops up on the grid" (via access to records such as the payment of utility bills) in enough detail for her and Bourne's next move to be predicted, which is possible because of a list of her previous addresses and associations filtered by their proximity to Paris.[38]

William Bogard, in an influential study inspired by a combination of works by Michel Foucault and Jean Baudrillard, has argued that surveillance is inextricably linked to its simulation—that the image of sophisticated mechanisms of total surveillance are as important to social control as is their actual functioning in reality—and it is no coincidence that the Bourne films exaggerate this activity. In the opening to *Ultimatum*, *Guardian* journalist Simon Ross's use of the word "Blackbriar" (an upgrade of the Treadstone program) while on his cell phone triggers a CIA alert that is relayed all the way to Vosen's New York office. In contrast to the historical reality of the failure of US intelligence to recognize threats, Vosen's action is swift and decisive, escalating quickly to the point at which he orders all the exits in the Waterloo train station to be blocked—now that Ross and Bourne are in there together—and that the "asset" (the euphemism for assassins such as Bourne) shoot to kill.[39] The political dimension of this fantasy is of naked state power, the taking of unilateral action on a foreign transport hub serving as an example of Vosen's claim that "Blackbriar" involves "no more red tape." It is also deeply connected to a technological imaginary in which smart technology is fetishized and exaggerated.[40] Thus, in the final film, video that appears to come straight from the viewfinder of a new asset chasing Bourne through a chaotic Athens in the grip of a mass riot is streamed back to CIA headquarters in real time. And while a scene such as this seems to fantasize to an extent that stretches the viewer's credulity to the limit, much of the panoptic representation in the films works precisely because it plays into the much deeper fantasy of surveillance that Bogard perceives in contemporary "reality." As Debbie Epstein and Deborah Lynn Steinberg thus argue, the utilization in the Waterloo station scene of actual footage from the station's CCTV system means that "while the fantasy of pursuit by embodied human agents remains fictional . . . , the fantasy of total surveillance by remote technology in *Ultimatum* is entirely real."[41]

Bourne's heroism, of course, lies in his ability to resist the grid, and much of his instinctive knowledge and behavior works to this end. He therefore mentions wiping surfaces down for fingerprints on several occasions, and his capacity to understand how he might be tracked informs much of his decision-making. The end of *Identity* is also premised on his success in this endeavor. In coming to see Conklin rather than Bourne as the heart of the problem, Ward Abbott (Conklin's immediate superior) activates one of the other Treadstone assets and has Conklin killed. When the success of this mission is reported to Abbott, again in the remote CIA headquarters location, his response—"Shut it down"—implies an ending in which both the pursuit of Bourne and the Treadstone project itself will be discontinued. Significantly, this is symbolically enacted when the agent to whom he gives the order shuts off all the computer terminals in the room. Bourne's freedom from authoritarian government is thus effected through a disabling of the networked devices of the grid.

What motivates and is implied by this resistance is a rejection of the identity imposed on him by the state. "Jason Bourne" is the CIA's object: trained to be an asset in the covert expansion of US imperialism. Each film culminates with one of Bourne's CIA superiors, blatantly coded as a father figure in all instances, attempting to reassert this master-slave relationship. When Bourne confronts Conklin in *Identity*, for instance, Conklin tells him that he is "a malfunctioning thirty million dollar weapon" and that "this is unacceptable, soldier." Dr. Albert Hirsch, the man behind the behavioral modifications implemented by the Treadstone project, takes a slightly different tack at the end of *Ultimatum*, saying to Bourne, "You made yourself. You chose right here to become Jason Bourne." Operating as the spiritual conclusion to the series, Bourne's response—"I remember everything. I'm no longer Jason Bourne"—works by abreaction to release him from his state-imposed identity and the behavior he has been driven to. Yet even prior to any understanding of who he is or was, Bourne runs from the state, an act represented in many ways as an instinctive preservation of the newfound subjectivity that rails against this imposition—a subjectivity that is awakened when he is unable to assassinate Nykwana Wombosi (the leader of an unnamed African state) as he plays with his children. Bourne's rebirth, as he comes out of the water at the start of the film, is thus as a morally aware and ethically sound individual who, in fact, instinctively acts to protect others such as Marie and the children of her friend Eamon.[42]

As Stephen Mulhall points out, whoever the figure played by Matt Damon is, his most entrenched belief is that "he is (above or before all) not Jason Bourne."[43] This innate resistance to the power of state-run agencies to name through identification positions him as the ideal embodiment of the autonomous modern subject. What makes this paradoxical is that Bourne completely lacks one of the capacities normally deemed fundamental to the creation of selfhood: memory. Bourne's memories of his training on the Treadstone project and of the subsequent missions he executed return only fleetingly across the series, and through a distorted lens, which emphasizes their insecurity. He also appears to have no memory of his childhood, of loved ones, or of where he grew up. Without a Humean chain of memories and associations, there is no self-identity for Bourne: he is incapable of telling a story about himself, the flashbacks in the film serving as a potent reminder of the very *disconnected* nature of his memory.

Where, then, does the overwhelmingly powerful sense of a "Bourne identity," which it could be argued sustains the series of films, come from? The most obvious answer is Bourne's exceptionality. Almost as soon as Bourne awakes in *Identity*, his surprising skill set is divulged, and it is on full display when he is stopped by two police officers while sleeping on a park bench in Zurich and instinctively disarms them and renders them unconscious through a series of specialized martial arts maneuvers. Crucially, Bourne is typically represented as seeing the world in quite a different way from almost everyone else, as is noticeable in his use of everyday objects. Bourne thus fights fellow assets with items such as a pen (*Identity*), a rolled-up magazine (*Supremacy*), and a book (*Ultimatum*), while many of his inquiries into the network of power in which he is entangled are completed by simple Google searches and the use of readily available information—a good example of this is the way he traces Landy's residence in Berlin simply by calling all the hotels in a tourist guide (*Supremacy*). Perfectly aligning with his lack of a past and his uncertain future, Bourne's skills position him as having a supreme mastery over the present. In his instantaneous perception of threat and innate understanding of space, Bourne is presented as inhabiting a ceaseless present that is traversed through the individual manipulation of objects: Bourne is the product of his object use. Yet while the embodied reflexes and instinctive skills that Bourne exhibits position him as exceptional and unique, they are also, of course, the result of the physical training and behavioral modification systematically imposed by the CIA. Bourne is thus at his most individual at the very moment that he is most obviously the state's object.

Intriguingly, many of Bourne's skills are also surveillant in themselves. His escape from the American embassy in *Identity* is possible because he masters both the layout of the building (absorbed almost instantly via a map he takes from a corridor wall) and the security communications system (achieved after he robs a guard of his radio). In numerous car chases around the unfamiliar streets of locations such as Paris, Moscow, and Goa, Bourne similarly has an uncanny ability to act as a human GPS. And it is his capacity to observe his surroundings in a remarkably perceptive manner that he finds the most uncanny of all the skills he has retained following his amnesia. While sitting in a motorway service station with Marie he tells her that:

> I can tell you the license plate numbers of all six cars outside. I can tell you that our waitress is left-handed and the guy sitting up at the counter weighs two hundred fifteen pounds and knows how to handle himself. I know the best place to look for a gun is the cab of the gray truck outside. . . . Now why would I know that? How can I know that and not know who I am?

Bourne's innate sense of how to manipulate the systems and technologies of surveillance is the prime means by which he turns the tables on each CIA agent who tracks him, turning them from hunter into prey. This often involves a cat-and-mouse game in which he first lures his opponents out so he can then place them under his surveillance. In *Identity*, for example, he arranges to meet Conklin on the Pont Neuf just so he can place an electronic tracking device on what he determines to be a CIA vehicle and thus trace the CIA back to their Paris safe house. *Supremacy* represents this facet of Bourne in even starker terms, as the drawing out is achieved through the identificatory technologies of the grid itself. When Bourne enters the Eurozone using his US—"Jason Bourne"—passport, he intentionally triggers an alert that causes the CIA agent who is questioning him to receive a call from Landy on his cell phone. Breaking out of the police station in which he is held, Bourne then uses his technical skills to mirror the agent's phone and thus listen in on his conversations with Landy, ultimately tracking her to Berlin and conducting his own phone conversation with her while watching her office through a rifle's viewfinder atop a nearby building. Indeed, Bourne's control of surveillance becomes so important that a scene in which he similarly watches Landy through her window both closes *Supremacy* and is replayed in its entirety in *Ultimatum*. The difference between the two is the context: in *Ultimatum*, Bourne is presented as enlisting Landy's help, which has, again, been prompted by "communication" through the intentional production of hits on the grid. Bourne's mastery is so complete by this point that Landy's ultimate tactic is simply to leave the office and walk the streets of Manhattan, telling a colleague that "if I'm out there he'll find me." The baffled looks of Landy and other agents as they examine the faceless cityscape in which Bourne is present but invisible during these scenes are clearly intended to represent a victory for the individual over institutional power. But it is also significant that Bourne's identity is formed partly through his own ability to surveil and through his self-imposed surveillance. Symbolically, this enfolds his portrayal as a character in the very system that he apparently seeks to resist, while his actions also revolve around an innate sense of his position within that system.

Memory, as an interior realm of interconnected impressions, is not the site of identity in the Bourne series. But there is another kind of memory at play in the films: namely, the traces and records of history that can be recollected through systems of identification. The grid, in this interpretation, is a map not just of movement but also of human motivation and agency. More explicitly, the memory of Bourne is held not inside his head but in the Treadstone files held by the CIA—files that are accessed through processes of identification. In *Supremacy*, Bourne's faked fingerprints are picked up by Landy's team, transferred by electronic scanner, and matched on an agency-networked computer. Landy is thus provided access to Bourne's file, turning her, as the files multiply in *Ultimatum*, into the benign keeper of Bourne's memory. This point is actually accentuated as Bourne's recollections begin to harden. In *Ultimatum*, Bourne tells Nicky Parsons that "I can see their faces. Everyone I ever killed. I just don't know their names." This scene is interspersed with images of Landy looking through the files of Bourne's missions, replete with the names of his targets. This extension of Bourne's mind is represented in various other ways—in *Supremacy*, he is seen recording his scattered thoughts, flashbacks, and dreams in a notebook in the hope that once on the page, they may begin to cohere—but it is strongly implied that it is only when he accesses his own file that he will truly know himself.[44] He does this through manipulating Vosen so as to produce replicas of his voice and fingerprint and thus gain access to the apparently unbreachable safe in his office. Such manipulations patently put technologies of identification at center stage at the precise moment in which Bourne apparently recovers his "real" identity. So Bourne is made complete not just as he traces his history but also as he characteristically brings his specialized skills to bear on a system of identification.

The identity that Bourne had before becoming Jason Bourne would, of course, have been subject to the same powers of identification and surveillance as every other citizen, but in the symbolic logic of the series it is strongly resonant of autonomy from institutional power. Landy gives Bourne this identity back to him in the phone conversation that closes *Supremacy* and is repeated in *Ultimatum*: Bourne's original name is David Webb, and his birthplace is Nixa, Missouri. If the films were exclusively guided by a liberal humanist ideal of individual autonomy and resistance to the state, then this should conclude the narrative. But the very banality of this information is indicative of its inability to sate the desire of either Bourne or the viewer. The audience members do not want Webb in Missouri. They want Bourne in Paris, London, and New York. This much is evident from the most recent film in the series, which is simply titled *Jason Bourne* and was advertised with the tagline, "You know his name."

The structural significance of this is that Bourne's representation entails a continual running from and resistance to being identified as Bourne, while it equally involves a deep engagement with the structures of power that named him Bourne in the first place. This tension is dramatized in a series of scenes. At the conclusion to *Ultimatum*, when Landy directs Bourne to the center in which he completed his Treadstone training (somewhat unbelievably located in central Manhattan), she addresses him as "David," and this scene is immediately followed by Bourne's confrontation with Hirsch, who repeatedly calls him "Jason." In a finale that includes numerous flashbacks, the viewer then sees the symbolic birth of Bourne take place when he hands "Webb's" military dog tags to Hirsch and shoots an unknown hooded victim, thus proving his capacity to act entirely on orders and with no concern for ethical justification. As Hirsch puts it, "You made yourself. You chose right here to become Jason Bourne," and this point is emphasized in *Jason Bourne* when CIA director Dewey attests that "you volunteered because you are Jason Bourne and not David Webb."

Hirsch and Dewey—and before them Conklin and Abbott—are father figures whose capacity to name and to speak the word of law is overcome. As if this representation were not blatant enough, the final film in the series accentuates the point further through the revelation that Webb's father, Richard, was the originator of the Treadstone project. Fulfilling the logic of Oedipal desire, all of these fathers end up dead. The one exception is Hirsch, who is symbolically castrated when Bourne refuses to kill him on the basis that "you don't deserve the star they give you on the wall in Langley." In keeping with this Freudian schema, the women who populate the series (except for Agent Heather Lee in *Jason Bourne*) are lovers/mothers who sacrifice themselves for

"Webb": indeed, Landy's revealing of his name and then addressing him as David acts as something of a rebirthing of Webb. Yet while he successfully grasps an agency that is enacted through the rejection of his agency-imposed identity—Bourne's confrontation with Hirsch ends with him claiming that "I'm no longer Jason Bourne"—he is never able to completely escape the Father, because a new one always appears to replace the previously deposed one. Significantly, Bourne is also never actually able to kill any of these fathers himself. Conklin is murdered, on Abbott's order, by an "asset"; Abbott himself commits suicide following the revelation of his corruption; and Dewey is shot by Lee. Bourne's inability to kill these figures superficially emphasizes the innate goodness of "Webb" and his attempt to find a way of being that lies outside the endless cycle of violence he has propagated (a position undermined by the body count that the films rack up outside of these scenes). On a deeper level, however, it is the specific position of these men as patriarchal authorities that renders Bourne impotent in their presence: like Hamlet, he is capable of extreme actions yet is made passive when faced by the figures who named him.

In what is a largely disappointing addition to the series, *Jason Bourne* does serve to notably emphasize two of the points this chapter has attempted to make in notable ways. First, in its numerous mentions of "Snowden" and in the creation of the Mark Zuckerberg–inspired social media CEO Aaron Kalloor, it brings the individual's acquiescence in systems of surveillance to the forefront and up to date. Indeed, Kalloor is introduced when making a speech about how the integration of a user's preferences will allow for a tailor-made experience (this is later shown to be in the service of a deal he has made with the CIA to give up the personal information of his users). Second, *Jason Bourne* lays bare the tension between the desire for autonomy and for subjection. Hirsch's analysis, given after the end of *Ultimatum* and then again in a file, is that "in my estimation Bourne will never find peace out in the cold," leading him to conjecture that "if played right Bourne could be brought back into the program." The idea of Bourne coming "back in" is, in the closing scenes of the film, rendered a distinct possibility; his conversation with Lee to this effect has been bolstered by a narrative that begins with his evident dissipation while living "off the grid" and even a scene in which he lingers longingly over his "Jason Bourne" passport. But of course, just as he can never entirely break away, Bourne can never fully return either. Rather, his fate is to continually run from the institutional power of the state to identify him in a manner that circles that very capacity. *Jason Bourne* thus ends like all the other films, with Bourne on the move, walking or swimming away from that which names him Bourne in a style and to a soundtrack that have become utterly distinctive of his "Bourne identity."[45] And his resonance as a character is entirely reliant on this very tension—on the fact that he exemplifies the contemporary subject whose desire for freedom is matched only by their desire to be named.

Notes

1. For a good example of this, see Craig Robertson, *The Passport in America: The History of a Document* (Oxford: Oxford University Press, 2010).

2. For examples of this, see chapter 2 in this collection and James Purdon, *Modernist Informatics: Literature, Information, and the State* (Oxford: Oxford University Press, 2016).

3. This phrase is adapted from Angus Calder, *The Myth of the Blitz* (London: Pimlico, 1991).

4. Neil Stammers, *Civil Liberties in Britain During the 2nd World War: A Political Study* (London: Croom Helm, 1983).

5. *The People*, editorial, 17 June 1951, RG 28 95, The National Archives, London, UK (hereafter TNA).

6. *The Times*, 27 June 1951, HO 45 25015, TNA.

7. "Judgment by the High Court of Justice, King's Bench Division, 26 June 1951," RG 28 95, TNA.

8. Jon Agar, "Modern Horrors: British Identity and Identity Cards," in *Documenting Individual Identity: The Development of State Practices in the Modern World*, ed. Jane Caplan and John Torpey (Princeton: Princeton University Press, 2001), 110.

9. Quoted in Edward Higgs, *Identifying the English: A History of Personal Identification 1500 to the Present* (London: Continuum, 2011), 202.

10. Ibid.

11. Ibid., 13.

12. David Barnard-Wills, *Surveillance and Identity: Discourse, Subjectivity and the State* (Farnham, UK: Ashgate, 2012), 18.

13. Ibid.; Alec J. Jeffreys, "The Man Behind the DNA Fingerprints: An Interview with Alec Jeffreys," *Investigative Genetics* 4, no. 21 (2013): 6.

14. "DNA Tests on Asylum Seekers Dubious in Law, Home Office Admits," *The Guardian*, 19 July 2018, https://www.theguardian.com/uk-news/2018/jul/19/dna-tests-asylum-seekers-dubious-home-office-admits.

15. Hayley Compton and Caroline Lowbridge, "How Familial DNA Trapped a Murderer for the First Time," BBC News, 22 September 2018, https://www.bbc.co.uk/news/uk-england-nottinghamshire-45561514. Allied to this is the prospect of what Michael Lynch describes as "reverse engineering a 'DNA photofit' of the appearance of a perpetrator of a crime from analysis of DNA left at the crime scene." Lynch et al., *Truth Machine: The Contentious History of DNA Fingerprinting* (Chicago: University of Chicago Press, 2008), 37.

16. Jeffreys, " Man Behind the DNA Fingerprints," 6.

17. Higgs, *Identifying the English*, 194.

18. Emilio Mordini and Andrew P. Rebera, "The Biometric Fetish," in *Identification and Registration Practices in Transnational Perspective: People, Papers and Practices*, ed. Ilsen About, James Brown, and Gayle Lonergan (London: Palgrave Macmillan, 2013); Katja Franko Aas, "'The Body Does Not Lie': Identity, Risk and Trust in Technoculture," *Crime, Media, Culture*

2 (2006): 145; Rebecca Gowland and Tim Thompson, *Human Identity and Identification* (Cambridge: Cambridge University Press, 2013), 89; Christian Parenti, *The Soft Cage: Surveillance from Slavery to the War on Terror* (New York: Basic Books, 2003), 5.

19. Kelly A. Gates, *Our Biometric Future: Facial Recognition Technology and the Culture of Surveillance* (New York: New York University Press, 2011).

20. William G. Staples, *Everyday Surveillance: Vigilance and Visibility in Postmodern Life* (Lanham, MD: Rowman and Littlefield, 2000).

21. See chapter 8 of this collection.

22. Irma van der Ploeg, "Written on the Body: Biometrics and Identity," *Computers and Society* (March 1999): 42.

23. Aas, "The Body Does Not Lie," 150.

24. Higgs, *Identifying the English*, 3.

25. David Lyon, *Surveillance Society: Monitoring Everyday Life* (Buckingham: Open University Press, 2001), 8.

26. Mark Poster, *The Second Media Age* (Cambridge, UK: Polity, 1995), 65.

27. Lyon, *Surveillance Society*, 15.

28. David Lyon, *The Electronic Eye* (Cambridge, UK: Polity, 1994), 4.

29. Kevin D. Haggerty and Richard V. Ericson, "The Surveillant Assemblage," *British Journal of Sociology* 51, no. 4 (2000): 617.

30. Clare Birchall, "Shareveillance: Subjectivity Between Open and Closed Data," *Big Data and Society* 3, no. 2 (2016): 2.

31. Lyon, *Electronic Eye*, 52.

32. Parenti, *Soft Cage*, 79.

33. Lyon, *Surveillance Society*, 75.

34. David Lyon, ed., *Surveillance as Social Sorting: Privacy, Risk and Digital Discrimination* (Abingdon, UK: Routledge, 2003).

35. For the sake of brevity, the films will be referred to as *Identity*, *Supremacy*, *Ultimatum*, and *Jason Bourne* for the rest of this chapter.

36. The amnesia thriller does have a history in fiction and became particularly prominent in the 1940s. See Victoria Stewart, *Narratives of Memory: British Writing of the 1940s* (Basingstoke, UK: Palgrave Macmillan, 2006), chap. 2.

37. For analysis along these lines, see Richard Pope, "Doing Justice: A Ritual-Psychoanalytic Approach to Postmodern Melodrama and a Certain Tendency of the Action Film," *Cinema Journal* 51, no. 2 (2012).

38. Klaus Dodds has written about the visual imagery of maps and mapping that takes place in these and other scenes in the Bourne films. Klaus Dodds, "Jason Bourne: Gender, Geopolitics, and Contemporary Representations of National Security," *Journal of Popular Film and Television* 38, no. 1 (2010): 31.

39. The stark contrast between this and the CIA's failure to recognize the significance of its own intelligence in relation to the 9/11 hijackers has been noted by critics. See Rand Richards Cooper, "Identity Crisis: The Bourne Ultimatum," *Commonweal*, 14 September 2007.

40. Debbie Epstein and Deborah Lynn Steinberg, "The Bourne Tragedy: Lost Subjects of the Bioconvergent Age," *Media Tropes* 3, no. 1 (2011): 96.

41. Ibid., 93.

42. This aspect of the series is very much related to a post-9/11 questioning of US foreign policy. See Vincent M. Gaine, "Remember Everything, Absolve Nothing: Working Through Trauma in the Bourne Trilogy," *Cinema Journal* 51, no. 1 (2011).

43. Stephen Mulhall, *The Self and Its Shadows: A Book of Essays on Individuality as Negation in Philosophy and the Arts* (Oxford: Oxford University Press, 2013), 98.

44. The influential idea of an extended mind comes from Andy Clark and David J. Chalmers, "The Extended Mind," *Analysis* 58, no. 1 (1998).

45. Epstein and Steinberg's interpretation has much in common with the one here, the point of departure being the issue of acquiescence and the source of subjection. So, they conclude that "Bourne's dilemma bespeaks a larger and conflicted cultural unconscious, one that acutely recognizes the totalizing tendencies of the bioconvergent imperative, and yet, at the same time, is intransigently persuaded by it." Epstein and Steinberg, "Bourne Tragedy," 109.

References

Aas, Katja Franko. "'The Body Does Not Lie': Identity, Risk and Trust in Technoculture." *Crime, Media, Culture* 2 (2006): 143–58.

Agar, Jon. "Modern Horrors: British Identity and Identity Cards." In *Documenting Individual Identity: The Development of State Practices in the Modern World*, edited by Jane Caplan and John Torpey, 101–20. Princeton: Princeton University Press, 2001.

Barnard-Wills, David. *Surveillance and Identity: Discourse, Subjectivity and the State*. Farnham, UK: Ashgate, 2012.

Birchall, Clare. "Shareveillance: Subjectivity Between Open and Closed Data." *Big Data and Society* 3, no. 2 (2016): 1–12.

Calder, Angus. *The Myth of the Blitz*. London: Pimlico, 1991.

Clark, Andy, and David J. Chalmers. "The Extended Mind." *Analysis* 58, no. 1 (1998): 7–19.

Cooper, Rand Richards. "Identity Crisis: *The Bourne Ultimatum*." *Commonweal*, 14 September 2007.

Dodds, Klaus. "Jason Bourne: Gender, Geopolitics, and Contemporary Representations of National Security." *Journal of Popular Film and Television* 38, no. 1 (2010): 21–33.

Epstein, Debbie, and Deborah Lynn Steinberg. "The Bourne Tragedy: Lost Subjects of the Bioconvergent Age." *Media Tropes* 3, no. 1 (2011): 89–112.

Gaine, Vincent M. "Remember Everything, Absolve Nothing: Working Through Trauma in the Bourne Trilogy." *Cinema Journal* 51, no. 1 (2011): 159–63.

Gates, Kelly A. *Our Biometric Future: Facial Recognition Technology and the Culture of Surveillance*. New York: New York University Press, 2011.

Gowland, Rebecca, and Tim Thompson. *Human Identity and Identification*. Cambridge: Cambridge University Press, 2013.

Haggerty, Kevin D., and Richard V. Ericson. "The Surveillant Assemblage." *British Journal of Sociology* 51, no. 4 (2000): 605–22.

Higgs, Edward. *Identifying the English: A History of Personal Identification 1500 to the Present*. London: Continuum, 2011.

Jeffreys, Alec J. "The Man Behind the DNA Fingerprints: An Interview with Alec Jeffreys." *Investigative Genetics* 4, no. 21 (2013): 1–7.

Lynch, Michael, Simon Cole, Ruth McNally, and Kathleen Jordan. *Truth Machine: The Contentious History of DNA Fingerprinting*. Chicago: University of Chicago Press, 2008.

Lyon, David. *The Electronic Eye*. Cambridge, UK: Polity, 1994.

———, ed. *Surveillance as Social Sorting: Privacy, Risk and Digital Discrimination*. Abingdon, UK: Routledge, 2003.

———. *Surveillance Society: Monitoring Everyday Life*. Buckingham: Open University Press, 2001.

Mordini, Emilio, and Andrew P. Rebera. "The Biometric Fetish." In *Identification and Registration Practices in Transnational Perspective: People, Papers and Practices*, edited by Ilsen About, James Brown, and Gayle Lonergan, 98–112. London: Palgrave Macmillan, 2013.

Mulhall, Stephen. *The Self and Its Shadows: A Book of Essays on Individuality as Negation in Philosophy and the Arts*. Oxford: Oxford University Press, 2013.

Parenti, Christian. *The Soft Cage: Surveillance from Slavery to the War on Terror*. New York: Basic Books, 2003.

Pope, Richard. "Doing Justice: A Ritual-Psychoanalytic Approach to Postmodern Melodrama and a Certain Tendency of the Action Film." *Cinema Journal* 51, no. 2 (2012): 113–36.

Poster, Mark. *The Second Media Age*. Cambridge, UK: Polity, 1995.

Purdon, James. *Modernist Informatics: Literature, Information, and the State*. Oxford: Oxford University Press, 2016.

Robertson, Craig. *The Passport in America: The History of a Document*. Oxford: Oxford University Press, 2010.

Stammers, Neil. *Civil Liberties in Britain During the 2nd World War: A Political Study*. London: Croom Helm, 1983.

Staples, William G. *Everyday Surveillance: Vigilance and Visibility in Postmodern Life*. Lanham, MD: Rowman and Littlefield, 2000.

Stewart, Victoria. *Narratives of Memory: British Writing of the 1940s*. Basingstoke, UK: Palgrave Macmillan, 2006.

van der Ploeg, Irma. "Written on the Body: Biometrics and Identity." *Computers and Society* (March 1999): 37–44.

10.
Identification and the "Intelligent City"

Dorothy Butchard

In J. G. Ballard's novel *Super-Cannes*, brochures for the futuristic corporate complex Eden-Olympia cast it as an "intelligent city," tracked and monitored by a pervasive network of cameras and security staff. This chapter explores the complex nexus of identification and surveillance in fictional portrayals of "intelligent" environments. My interest in literary representations of intelligent spaces is a deliberate response to contemporary trends toward "smart" environments in city planning and the consequent intersection of networking technologies with urban public infrastructure. Broadly understood as a space in which "information technology is combined with infrastructure, architecture, everyday objects, and even our bodies to address social, economic and environmental problems," smart cities are frequently presented in utopian terms as the bright future of high-density urban living.[1] But although there are reasons to feel hopeful regarding intelligent infrastructure's capacity to contribute to cleaner, greener, more efficient environments for large populations, it is worth pausing to consider the ramifications of a system reliant on sustained monitoring in "a city activated at millions of points."[2] Reading speculative novels by J. G. Ballard and Nicola Barker through the lens of scholarly debates about the twenty-first-century smart city, I show how understandings of identification are brought into sharp relief where citizens' lives meet intelligent spaces.

The works of fiction I discuss in this chapter use fictional scenarios to contemplate the attractions and pressures of living in an intelligent city. J. G. Ballard's *Super-Cannes* (2000) and Nicola Barker's *H(A)PPY* (2017) offer provocative test cases for exploring the art of identification as it becomes entwined with smart systems. As Matthew Jewell notes, cities have always been "spaces of conflict, community, power and politics that involve complex negotiations of space, convention and order," and the intelligent city is no exception.³ The novels I discuss here explore those "complex negotiations" in the context of hyperconnected environments and intensive surveillance systems. Part business park, part "ideas laboratory," the pseudocity of Eden-Olympia in *Super-Cannes* is a carefully designed space that aspires to maintain "convention and order" through a combination of perpetual monitoring and prescribed violence. In *H(A)PPY*, Barker presents the exaggerated apotheosis of a "smart city," in which citizens' emotions are seamlessly embedded within social infrastructures and residents are encouraged to live in a state of "perfected" neutrality. Both novels offer contentious and perceptive interpretations of a social space whose continuing existence is predicated on constant acts of identification and regulation.

Although scholarly discussions of smart cities are ostensibly focused on improving efficiency, infrastructure, and living standards for urban citizens, such commentaries do not usually admit the personal ramifications of life within hyperregulated spaces. For example, Marcus Foth observed in 2016 that a "smart city plan" published that year "uses language that conveys a limited role for people in cities: they live, work and consume."[4] Instead, the real-world occupants of smart cities tend to be abstracted into information "nodes" valued for the data they can supply to facilitate the city's smooth functionality. By contrast, the novels I discuss in this chapter present distinctive protagonists whose relationship with the high-tech environments they occupy is fraught with conflicting emotions. Paul Sinclair, narrator of *Super-Cannes*, is both enticed and appalled by the carefully curated perfection of the intelligent city of Eden-Olympia. In *H(A)PPY*, the emotions of Mira A are totally entwined with the spaces she occupies in "The System," as she veers between determination to fit in with her sensor-driven environment and involuntary impulses to disregard its unspoken rules. Both novels use their protagonists' oscillation between admiration and repulsion to explore a complex nexus between personal identity, technologized monitoring practices, and automated modes of identification.

Identification as Preservation

Super-Cannes introduces the newly built complex of Eden-Olympia as a glittering model for an imagined city. With its carefully cultivated public spaces, perfectly aligned car parks, and inscrutably screen-like buildings, Eden-Olympia's aesthetics are more immediately akin to a gigantic business park than to a speculative city. Nevertheless, it has lofty aspirations. Early in the novel, Eden-Olympia's in-house psychiatrist Wilder Penrose describes it with characteristic hyperbole as a "huge experiment in how to hothouse the future."[5] Introducing the park to new arrivals Jane and Paul Sinclair, Penrose describes it as a unique template:

> "Eden-Olympia isn't just another business park. We're an ideas laboratory for the new millennium."
> "The 'intelligent' city? I've read the brochure."
> "Good. I helped to write it."[6]

Penrose's concept of Eden-Olympia as an "ideas laboratory" chimes with the kind of language used to describe the development of smart cities in the twenty-first century. Descriptions of nascent smart spaces across the world regularly use terms such as "experiment" or "laboratory"; Hans Schaffers, Carlo Ratti, and Nicos Komninos could be quoting Penrose when they describe "an urban innovation ecosystem, a living laboratory acting as agent of change."[7] Ballard's own references to Sophia-Antipolis, the business park cited as an inspiration for *Super-Cannes*, rework these preoccupations with the idea of a "future" space in less glowing terms; Ballard's reference to "the way I see the future—endless gated communities, high technology taking the place of human relationships" is tempered by the prospect of "a deep unconscious boredom."[8] Reading *Super-Cannes* in light of this rhetoric illustrates how claims to progressive futurity intertwine with concerns about infrastructures dedicated to monitoring and regulating their citizens.[9]

The hyperbolic claims about Eden-Olympia as an "ideas laboratory" in *Super-Cannes* simultaneously examine and critique the allure of the intelligent city for its population of privileged executives. In an interview, Ballard muses that "most of my so-called dystopias are rather pleasant places to live,"[10] and Frida Beckman considers this a particular feature of Ballard's later novels, as "dystopian wastelands and natural disasters" give way to "spaces of perfection."[11] Paul Sinclair, narrator of *Super-Cannes*, describes an improbably perfected space, "as if the entire business park were a mirage, a virtual city conjured into the pine-scented air like a son-et-lumière vision of a new Versailles."[12] The pleasantly sanitized physical attributes of Eden-Olympia are conveyed via conventional hallmarks of luxury; even the clinic has a reception area comparable to "the sun deck of a cruise liner."[13] Lee Rozelle situates these accounts in the specific era of the novel's publication, observing that *Super-Cannes* "depicts utopian communities and imaginations pulling from twentieth-century urban landscape into millennial corporate possibilities in city design."[14] Descriptions of Eden-Olympia's architecture are steeped in references to urban visionaries and efforts to develop purpose-built "utopian" spaces.

Ballard's emphasis on clean lines and efficiency evoke a rhetoric of innovative city-building and progressive urban planning that was established long before the advent of smart cities at the turn of the millennium. Given Ballard's career-long interest in dystopian cityscapes, it is perhaps unsurprising that *Super-Cannes* is alert to the mixed legacy of nineteenth- and twentieth-century architectural innovation. Paul interprets brochure images of Eden-Olympia as "a humane version of Corbusier's radiant city," later returning to this comparison as he gazes at "lines of office buildings rising from the park like megaliths of the future."[15] The account of Eden-Olympia in *Super-Cannes* consciously echoes Le Corbusier's unfulfillable promise of a new model for modern living. Penrose proudly refers to his prototype for "Eden II" as "the new Europe" and responds to Paul's ironic quip that this is not Winthrop's "City on a Hill" with a dizzying pledge to build "a hundred cities on a hundred hills."[16] In particular, Eden-Olympia's utopian ambitions evoke Corbusier's insistence that architecture should demand "health, logic, daring, harmony, [and] perfection" while also exposing the more sinister possibilities of a space designed with these aims in mind.[17] Just as Yevgeny Zamyatin's *We* presented a gleaming city of "immutably straight streets" where "everything was in its place, so simple, normal, legitimate" in order to explore the dangers of techno-authoritarian repression, *Super-Cannes* quickly exposes the lie of Eden-Olympia's harmonious exterior.[18] To achieve this, Ballard explores how Eden-Olympia's apparently utopian qualities are fundamentally entwined with the surveillance practices that underpin its infrastructure.

Eden-Olympia depends on several kinds of surveillance to promote the security and well-being of its privileged citizens, with familiar hallmarks of twentieth-century surveillance assemblages including CCTV, identity cards, and private security guards. *Super-Cannes* also explores more subtly intrusive models of monitoring and identification. On Jane and Paul Sinclair's initial tour of Eden-Olympia, Penrose half jokes that he recommends a heart attack or plane crash "once you're settled in," since "the paramedics will know everything about you—blood groups, clotting factors, attention-deficit disorders."[19] The phrase "once you're settled in" continues an enduring theme in Ballard's writing, announcing the violent crash as a means to disrupt the boredom of a "settled" existence and to "lay bare the psychopathologies of everyday life."[20] It also draws attention to this pseudocity's gathering of intelligence, demonstrating Eden-Olympia's ability to connect and monitor its citizens based on data accumulated as they "settle in" to its environment. The lighthearted reference to a medical team who will "know everything about you" implies that to live in this space involves complete absorption into its insulating environment, where threats to bodily security can be dealt with swiftly and easily. Once a citizen has become "settled," the information accrued during the person's residence in the intelligent space offers a promise of total resolution—in this example, the capacity to reverse physical frailty or life-threatening injury. As Paul notes sardonically later in the novel, Eden-Olympia aims to promise "if not immortality, then perpetually-monitored health."[21] The aspirational intelligence of Eden-Olympia represents a promise of well-being enabled by people's willing integration into a larger superstructure. It is identification as preservation.

The intelligent automation of health care and security is a crucial component for Ballard's scrutiny in *Super-Cannes* of the self-styled "ideas laboratory for the new millennium." This is emphasized via the role of Paul's wife, Jane, who joins the Eden-Olympia medical team. Critical discussions of *Super-Cannes* have tended to overlook the figure of Jane, whose professional activities are largely ignored by her narrating husband except insofar as they remove her attentions from him. Nevertheless, brief references to Jane's work at Eden-Olympia are highly instructive for thinking about the modes of identification examined in Ballard's novel. The project Jane is working on aims to monitor and evaluate the health of citizens:

> "Every morning when they get up people will dial the clinic and log in their health data: pulse, blood-pressure, weight and so on. One prick of the finger on a small scanner and the computers here will analyse everything: liver enzymes, cholesterol, prostate markers, the lot."
>
> "Alcohol levels, recreational drugs . . . ?"
>
> "Everything. It's so totalitarian only Eden-Olympia could even think about it and not realize what it means."[22]

This is privatized biosurveillance in the sense of a process that *"systematically collects and analyzes data for the purpose of detecting cases of disease."*²³ The claimed capacity to "analyse everything" echoes Penrose's assertion that "the paramedics will know everything about you" while explicitly shifting responsibility for this knowledge to the system that gathers and stores data. The repetition of "everything" reiterates Penrose's aspiration to all-encompassing knowledge, as does the insistence that this will become a necessary part of citizens' daily routine, to be shared "every morning." As the novel unfolds, Jane's interest in the project's potential appears to outweigh her early skepticism at its "totalitarian" tone: "The work here is so interesting. We may be on to something with these self-diagnostic kits. The first hint of liver disease and diabetes, prostate cancer. . . . You don't realize what a single drop of blood can say about you."²⁴ The emphasis on "self-diagnostic kits" and the revelatory potential of a "single drop of blood" combines the promise to identify truths latent in previously undiagnosed bodies with the need for citizens' compliance in these practices of protective identification.

Like Ballard in *Super-Cannes*, Barker presents the capacity to promote and maintain physical well-being as a key feature of intelligent environments, which are designed to be solicitous of and responsive to citizens' physical and emotional health. Eden-Olympia's approximation of paradise has been designed precisely as a "space of perfection" for a high-functioning executive class, and *H(A)PPY* goes further in linking the perfected status of citizens with their environment's claims to promote symbiotic well-being. "We are In Balance," the narrator Mira A explains. "The System expresses us perfectly, and we express it perfectly."²⁵ *H(A)PPY* centers on a large community of "perfected" individuals—known as "The Young"—who exist in a purpose-built environment:

> We were perfected. . . . Before, there was filth and it corrupted us. After, there was freshness. There was the smell of newly cut grass. Everything shone. They made us feel innocent again. No—*no*. They made us Innocent again.²⁶

Echoing the "pine-scented air" of Eden-Olympia's artificially constructed pseudoparadise, the "perfected" environment occupied by "The Young" in *H(A)PPY* is described using pastoral imagery. Freshness is a recurring trope in the novel, repeatedly associated with purity, neutrality, and innocence. When

Mira A feels loved by her community, she describes a state of harmony that feels "like a soft shower of apple blossom"; other instances of this harmony include "a soft, pale glow in the sky" and "evening air . . . scented with the reinvigorating aroma of crushed rosemary."[27] But Barker consistently sets this pastoral imagery against reminders of the artificiality of Mira's surroundings, a "perfected" space monitored by a complex system of sensors and self-regulation. *H(A)PPY* imagines a society built for and predicated on a complex system of sensor-based monitoring and self-surveillance.

H(A)PPY is dominated by amplified versions of contemporary smart devices, which are depicted as essential to the maintenance of a fully responsive environment. Everything Mira and her fellow citizens do, say, or feel is identifiable and quantifiable through a display on a system known as "The Sensor." With its aural pun on "censor," this system represents a more pervasive form of monitoring than any imagined in *Super-Cannes*. Everything in Mira's world feeds into The Sensor, which then responds with automated adjustments. For example, clothing is designed to track and adapt to citizens' bodily conditions:

> All our fabrics are intelligent now. We grow them in laboratories. Our fabrics are self-cleaning and self-maintaining and they interact with our bodies to gauge things like size, density and temperature according to the specifics of the conditions in which we find ourselves. Our fabrics—our shoes—are alive.[28]

This represents a far more advanced intelligence than does Ballard's depiction in *Super-Cannes*, in which biomedical data must still be actively volunteered by citizens who submit to "the endless check-ups Eden-Olympia arranges."[29] In *H(A)PPY*, intelligent fabrics are part of a fully integrated system dedicated to automated monitoring and response. The fabrics reflect the aims of sensor-based technology implicit in twenty-first-century development of "wearables" to enable "remote health monitoring of patients."[30] Their seamless protection demonstrates the appeal of data collection via ubiquitous devices, offering the capacity to monitor and regulate human conditions in a way that is both unobtrusive and reassuring, as in real-life aspirational twenty-first-century technologies "at the fringe of our awareness" that exert significant control yet appear "so mundane, that we hardly notice."[31] In moments of distress or dissent, however, the fabrics become opponents: when Mira A claws at her chest in pain, "the fabric prohibits [her]."[32] They represent a tension between protection and prevention that is inherent in automated systems designed for bodily, if not emotional, preservation.

Barker's novel introduces the risks of total integration with an intelligent environment. Mira is equipped to fit The System in ways that make her unsuitable for a nonsystem; when she later faces ejection into the oppositional wilderness space of "The Unknown," she learns that implants installed to monitor and regulate her behavior "will be rejected by [her] body out there."[33] As she slips outside the jurisdiction of The System, Mira finds her clothing has become "sluggish."[34] Having "expected the intelligent fabrics to protect [her]," she discovers her skin exposed and worries that others may "infect her with something."[35] The chaos of The Unknown is indicated by its incapacity to protect its residents from such threats to bodily well-being; an intermediary warns that "we do not have medicine like The Young. You are better off staying here."[36] But the perfected, protected state of "The Young" is reliant on their continuing compliance with the hyperregulation of their environment, and this willingness to comply with surveillant assemblages is an essential factor in the preservation of intelligent environments in both *H(A)PPY* and *Super-Cannes*.

Self-Monitoring and Internalized Surveillance

As the earlier examples have begun to demonstrate, life in an intelligent city requires compliance with a technological infrastructure designed to trace, connect, and evaluate its citizens. Jathan Sadowski and Frank Pasquale describe this effect when they argue that "acting as human information nodes in the urban network is becoming another civic and economic duty that smart city dwellers are expected to perform."[37] This emphasis on the responsibility of urban citizens is by no means a new phenomenon. In *The Rule of Freedom*, Patrick Joyce's account of nineteenth-century urban planning captures the aim to create "a city of wide streets, straight lines, improved visibility..., cleaner and better lit than ever before" whose citizens are "responsible and therefore self-monitoring."[38] This assumption of responsibility and self-monitoring becomes exponentially intensified in the intelligent spaces depicted by Ballard and Barker. In these novels, as in real-life smart environments, citizens move through systems that cater to their needs only insofar as they agree to contribute to the city-space's smooth functionality, with their compliance monitored via the collection of personal data or their interactions with automated sensors. Both novels therefore explore what it means to become a "human information node," envisaging spaces that are actively designed to insist on citizens' responsibility and active self-monitoring.

Mark Poster's concept of the "superpanopticon" is a useful reference for considering the ramifications of identification in the kinds of spaces envisioned by smart-city developers across the world. More than a decade before the development of sensor-driven infrastructures or the commercial production of wearables, Poster characterized the networked databases of emergent information technologies as "a gigantic and sleek operation... whose political force of surveillance is occluded in the willing participation of the victim."[39] Poster's account of the superpanopticon envisages databased selves imbricated in a global network:

> The phone cables and electric circuitry that minutely crisscross and envelop our world are the extremities of the superpanopticon, transforming our acts into an extensive discourse of surveillance, our private behaviors into public announcements, our individual deeds into collective language. Individuals are plugged into the circuits of their own panoptic control.[40]

In *H(A)PPY*, Mira and her peers are willing participants in the elision of "private behaviors into public announcements" in an intensively networked space. This participation is made literal in the figure of "The Graph," a universal display that uses color coding to regulate emotions. The Graph "shows us how In Balance we are: as a person (our physical and mental health), as a small community (a community of skills, a community of friends, a community of consumers, a community of thought) and as a broader society—as a race, as a planet, as a galaxy."[41] Mira's expanding list—from person to community and all the way out to galaxy—visualizes Poster's impression of the reconfiguration of "individual deeds into collective language."[42] The emphasis on balance, community, and collective responsibility transforms the familiar language of civic duty into a superpanoptic network of plugged-in selves. All graphs are available to all individuals, and the result is a system predicated on voluntary neutrality.

In the superpanoptic spaces of *H(A)PPY*, any "Excess of Emotion" is treated as a crisis, and the intelligent environment is designed to ensure that citizens identify and regulate their own behavior before it can become "excessive." When she becomes increasingly unable to maintain the neutral emotions required by The System, Mira battles against her own volatile self-perception. In one of the novel's most startling passages, Mira stumbles into a music group she has joined, only to discover her peers all watching the excesses of emotion unfolding on her Graph:

> They were actually . . . poring over the consolidated footage of *intimate details of my day*. Yes. They were watching me. They were studying me. Mira A. The large screen was divided into several parts. In one part I was running on the Power Spot and I was plainly exhausted. My Graph was pinkening quite dramatically because of the excessive and—quite frankly—unnatural levels of effort I was expending. I shuddered at the sight of it. My behaviour was silly. It was inappropriate. It was utterly counterproductive. (How had I not realised? How had this plain truth escaped me? This activity was bogus and completely self-defeating!)[43]

As Mira watches her peers watching her, she actively reinterprets her own behavior. In their horrified response, Mira sees how her own actions have deviated from the expected "balance," prompting another emotional reaction that she immediately tries to regulate. Tim Jordan speculates that "the fear of the Superpanopticon is of a world in which deviance from the norm is already guilt."[44] We see this anxiety at work in Mira's distress at witnessing her "unnatural" actions through the eyes of others and in her subsequent self-flagellation at the "inappropriate" excesses of her exercising. Barker thematizes this concern throughout *H(A)PPY*, which explores how a system predicated on self-regulation in a virtuous community may lead to all expressions of individuality being pathologized. In Mira's case, her fellow musicians are quick to shut down the situation, lest it "infect" their own Graphs. "If both of these Communities become implicated in your purpling it will critically affect the Balance in at least two of my activities," one of her fellow musicians explains. "Two of my Graphs—my Communities—at the very least, will be implicated. So calm down. Please calm down."[45] Throughout the novel, intelligent citizenship involves the curation of others' neutrality as well as one's own.

Acquiescing to her role as "information node," Mira does not question the systems by which her actions have been recorded and accessed. Instead, her immediate response to the revelation of "intimate details" is to chastise herself and endeavor to fit within the limits of regulated behavior. When Mira tries to modify her thoughts and behavior in anticipation of surveillance, she acknowledges what Michel Foucault defines as "a state of conscious and permanent visibility that assures the automatic functioning of power."[46] Mira's determined effort to appear suitable to an imagined audience of "Clean, Young eyes, to Neuro-Typical eyes" envisions an extreme end point for Foucault's concept of panopticism, in which "constant pressure acts even before the offences, mistakes, or crimes have been committed."[47] For Clare Birchall, this panoptic end point can be understood in terms of the contemporary sharing of information "not as a conscious and conscientious act but as a key component of contemporary data subjectivity," incentivized by pragmatism and privilege.[48]

> Citizens share data with a proprietary agent in exchange for the *privileges that come with citizenship*. We might, that is, consciously or unconsciously, explicitly or implicitly consider the collection of our GPS data or phone metadata *a fair price to pay* for the freedoms, benefits, and protections that come with owning a British (or Australian, German, American, etc.) passport. . . . Users of social media and search engines are familiar with making trade-offs between services they want and acquiescence to data collection.[49]

The result is a system predicated on compliance, in which self-surveillance and the monitoring of others is a way to ensure continued status and privilege, and active citizenship is motivated by the fear of falling out of privileged communities. Mira's encounter with The Graph presents an exaggerated apotheosis of Birchall's understanding of contemporary data subjectivity. In *H(A)PPY*, the trade-off is never as concrete as sharing data in exchange for a specific service. Instead, Mira acquiesces to a state of constant "shareveillance" because it is integral to her place within her various communities. This is no longer conceived of in terms of an exchange or price to pay but as a state in which each individual accepts self-surveillance as a prerequisite for maintaining the system in which they continue living.

If *H(A)PPY* offers a vision of an entirely integrated surveillance society, *Super-Cannes* depicts an embryonic prototype of the surveillant potential of an intelligent city. Eden-Olympia is a flawed and incomplete precursor to the automated surveillance assemblages represented in Barker's novel. A conversation between Paul Sinclair and security guard Frank Halder reveals the limitations of a surveillance assemblage still dependent on human processing; the scale of CCTV capture in Eden-Olympia means that "scanning the tapes . . . will take a great many hours of overtime."[50] This situates the concerns of *Super-Cannes* firmly within a late twentieth-century understanding of CCTV's role as an agent of deterrence, linked to "fear of crime and perceptions of insecurity as much as with actual occurrence."[51] In a much-quoted passage from the novel, Paul Sinclair reflects on what he sees as a triumph of "invisible infrastructure" in Eden-Olympia:

> An invisible infrastructure took the place of traditional civic virtues. At Eden-Olympia there were no parking problems, no fears of burglars or purse-snatchers, no rapes or muggings. . . . Civility and polity were designed into Eden-Olympia, in the same way that mathematics, aesthetics and an entire geopolitical world-view were designed into the Parthenon and the Boeing 747. Representative democracy had been replaced by the surveillance camera and the private police force.[52]

Super-Cannes takes the principle of deterrence that underpins CCTV and extends it beyond the physical infrastructure of the city to the social fabric of citizenship, in which "honesty is a designed-in feature, along with free parking and clean air."[53] Early in the novel, Wilder Penrose argues that "the important thing is that the residents of Eden-Olympia think they're policing themselves," a statement that anticipates the self-regulation we find in *H(A)PPY* and the shareveillance that Birchall identifies as the price of citizenship.[54] As Paul Sinclair discovers, anyone who does not contribute directly to the "invisible infrastructure" is liable to either conversion or violent removal, while those who comply reap rewards of wealth, luxury, comfort, and security.

The smooth running of Eden-Olympia is a continual work in progress, and its citizens are expected to contribute. Jewell argues that "smart city initiatives seek to impose upon urban space a technology-mediated transformation of seamless efficiency," and this insistence is evident wherever willingness to promote the intelligent city has become an essential criterion for continuing to exist within it.[55] For much of *Super-Cannes*, Paul appears to take pleasure in his own unproductive state, contrasting his broken leg and disheveled appearance with the smooth functionality of his surroundings. But even Paul is reinterpreted as "useful" to Eden-Olympia, as he is informed late in the novel that he has been used as a test subject for understanding the pathologies that previously led to a series of murders. The insistence on citizens' willingness to act as "human information nodes" leads to an environment that overtly rejects, belittles, or excises those who do not contribute to the continuing promotion of technological progress and commercial influence.

In discussing smart cities, Martijn de Waal, Michiel de Lange, and Matthijs Bouw observe tensions between the ideal of a smart environment and its lived reality, arguing that "cities are also sites of contestation, where different groups of citizens may have different preferences or economic interests, and where existing power relations do not all of a sudden disappear with the emergence of collaborative digital media platforms."[56] This recognition is essential to the narrative tensions in *H(A)PPY* and *Super-Cannes*, whose advanced monitoring systems repeatedly fail to prevent these spaces from becoming "sites of contestation." In *Super-Cannes,* Eden-Olympia is dogged by multiple interferences with its claims of security and efficiency, while The System in *H(A)PPY* ultimately cannot contain Mira's inadvertent rebellions. In both cases, glitchy individuals appear to threaten both system and infrastructure, illustrating conflicts between an idealized concept of highly regulated, luxury urban spaces and a continuing recognition that "risk, uncertainty, exposure, self-consciousness and vulnerability are inevitable and unresolvable aspects of our being in city spaces."[57] In this concluding section, I consider how imagery of infection is used to represent uncertainties and failures—and the efforts to inoculate the intelligent city against such risks.

Despite its rarefied air and efforts at regulation, Eden-Olympia is an experimental space, subject to the kinds of challenges described in Myria Georgiou's account of high-density living as "struggles for power, control and ownership . . . reflected and shaped through the intense (mediated) meetings of people, technologies and places."[58] Concerns about potential threats to the "spaces of perfection" are evident in both novels, though they are treated in strikingly different ways. Even before the narrative begins, the failings of Eden-Olympia's efforts to reduce "struggles for power" have been exposed by a series of violent murders targeting the elite of *Super-Cannes*; this failure of the regulated environment haunts the novel as a mystery its narrator seeks to unravel. By contrast, Mira's accounts of The System at first suggest that struggles for power and ownership have been eliminated, but Mira chafes against the behavioral and linguistic restraints of her cocooned world. In *H(A)PPY,* the oscillation between pastoral idyll and unregulated peril is underpinned by recurring references to an "unperfected" other world—The Unknown, troubled by "plague . . . and turmoil and hunger."[59] In *Super-Cannes,* the equivalent of The Unknown is found in the nearby slums of Cannes, Nice, and La

Bocca. Both novels treat these uncontrolled spaces as figurative antitheses to the highly regulated environments occupied by their protagonists.

Throughout *H(A)PPY*, Mira's perceived threats to the balance of The System are described in terms of infection and bodily imperfection—she is "impure," "leaking" and subject to "an itch" and "a minor endocrine fail."[60] As Mira's emotional distress intensifies, it is mirrored in her body; she becomes "slow to heal," disturbed by "conflict at a cellular level."[61] Her compliance is portrayed as essential to her environment's continuity; her failure to comply is figured as a poisonous infection. Another character speculates that these failings may be genetically embedded and are likely to have rendered her subject to intense observation.

> It's possible that the Technicians have detected a flaw in your genetic make-up—something ancient and unresolved—and they're doubtful that they can fully control it. They will have been watching you for years, even decades, waiting for it to develop. It will be something tiny but irresistible. An oscillation. An urge.[62]

These acts of biological identification recur in Mira A's narrative. Her wayward body and correspondingly wandering mind are subject to constant observation. This monitoring leads to modification: when she begs for medical treatment, she is promised curative "chemicals" and adjustment to her "Oracular Devices" and is fitted with "clamps" that will "place a measure of pressure ... on to the casings of the brain."[63] The adjustments to Mira's brain recall Paul's sardonic joke that the brains of Eden-Olympia's citizens will "soon need a false ceiling to make room for the ducting demanded by our 'intelligent' lifestyle."[64] Paul's hyperbolic imagery displaces such modification to a future possibility, with brains "ducted" in a vision reminiscent of the grounds of a nearby residence "so crammed with electronic ducting that no roots could prosper."[65] In Barker's vision of a more distant future, the flippant analogy has become a lived reality.

The efforts to anticipate and control Mira's troublesome rebellions in *H(A)PPY* reflect a broader ambition to develop urban environments with the capacity to anticipate and forestall potential damage and disruption. Unable to constrain herself to the neutral language required by The System, Mira is warned that her involuntary deviations may pose "a serious threat to The System"; they could "poison," "pollute," and "unbalance" its functions.[66] Intelligent spaces are increasingly designed to anticipate, assimilate, and/or divert potential sources of tension or conflict. By reimagining Rosa Parks's famous stance against racial segregation in the environment of a smart city, Evgeny Morozov illustrates how smart environments might aim to inoculate cities against individual sources of dissent or disruption.

> Imagine that Parks is riding one of the smart buses of the near future. Equipped with sensors that know how many passengers are waiting at the nearest stop, the bus can calculate the exact number of African Americans it can transport without triggering conflict; those passengers who won't be able to board or find a seat are sent polite text messages informing them of future pickups. . . . Those passengers most likely to cause tension on board are simply denied entry.[67]

Morozov's example demonstrates how an intelligent environment can displace agency from individuals to a dispersed system—for example, in the use of sensors whose role is managerial rather than aggressive but whose capacity to identify and regulate a space plays an active role in the effort to neutralize and displace potential sources of conflict. This displacement effect is further evident in the imagined use of "polite text messages" to refuse access, with the promise of "future pickups" less likely to prompt disturbance. Finally, the phrase "simply denied entry" draws attention to the overriding principle at the heart of each of the speculative intelligent environments I am discussing here: the use of technologies and infrastructure to segregate and dissociate potential troublemakers—or indeed, anyone who, like Mira in *H(A)PPY*, might "cause tension" or fail to comply with the established structural regime.

Morozov's near-future smart scenario opens up questions of access and accessibility, highlighting the power of the denial of entry within urban spaces. Morozov's reimagination of the Parks protest draws attention to the seamless integration of racial profiling with the concept of the gateway, in this case via automated denial of access to specific services based on the identification of skin color. In the scenario Morozov describes, the ease of identification based on categorization—the automated detection and refusal of African American passengers—has become part of a totalizing system equipped to automatically permit or deny entry. A lower-tech version of this profiling is evident in *Super-Cannes*, in which darker skin tones are instinctively associated with immigrant Arab populations. Penrose bemoans the many ways in which Eden-Olympia remains permeable to an influx of uninvited presences:

> These gewgaw men get in anywhere. Somehow they've bypassed the idea of progress. Dig a hundred-foot moat around the Montparnasse tower and they'd be up on the top deck in three minutes.[68]

Penrose's statement, including the contemptuous slur "gewgaw men," channels the racist tones of technocapitalist language. When he dismisses the Arab traders as bypassing "the idea of progress," Penrose follows what Alondra Nelson calls a well-established "scenario" in which "blackness gets constructed as always oppositional to technologically driven chronicles of progress."[69] *Super-Cannes* is built around such moments, which appear to expose inherent flaws in Eden-Olympia's systems for security and preservation of its executive systems. But whereas Nelson notes that Blackness is commonly viewed as "the anti-avatar of digital life," Paul reinterprets the traders as transformed by Eden-Olympia's sleekly futuristic aspirations; their faces "had a silvered polish, as if a local biotechnology firm had reworked their genes into the age of e-mail and the intelsat."[70] This incorporation implies an involuntary modification akin to the ducting of brains described earlier. In doing so, it suggests the excessive reach of the intelligent city, capable of transfiguring temporary visitors into the shape and frame of its futuristic environment.

The "paranoid new world" of *Super-Cannes* is troubled by the anticipation of implied infection; Penrose claims that "the health of Eden-Olympia is under constant threat."[71] *Super-Cannes* depicts an effort to integrate and legitimize the kinds of disruptive impulses Mira battles to suppress in *H(A)PPY*. Whereas Mira is ultimately exiled from The System for "Excesses of Emotion," Eden-Olympia's highest-ranking citizens are encouraged to channel violent excesses into personal well-being and workplace efficiency. Faced with an array of ailments among the executive denizens of Eden-Olympia—"respiratory complaints," "bladder infections and abscessed gums," "opportunist fever," "malaise," and "chronic fatigue"—Penrose opts to "prescribe" sessions of transgressive violence. He describes these violent attacks as "a controlled and supervised madness":[72]

> Here at Eden-Olympia we're setting out the blueprint for an infinitely more enlightened community. A controlled psychopathy is a way of resocializing people and tribalizing them into mutually supportive groups.[73]

In Penrose's diagnosis, the physical ailments are manifestations of "malaise" resulting from the absence of "personal morality" in an environment so secure, so intelligent, that "the moral order is engineered into their lives along with the speed limits and the security systems."[74] Ostrowidzki describes Eden-Olympia as "a neoliberal, neo-imperial, and neofascist Utopia of the Far Right,"[75] and the "therapy groups" represent Penrose's attempt to co-opt and reintegrate threats from what he calls the "contingent world" as he seeks to maintain influence over actions and responses that are unpredictable and not easily controlled.[76]

Penrose's attempt to use "controlled" violence in pursuit of a "richer, saner, more fulfilled" citizenship is a key example of Eden-Olympia's dependence on the use of advanced technologies to inoculate against glitches and disruptions. The so-called therapy sessions prescribed by Penrose are tracked and recorded by the participants themselves, so that handheld video cameras and camcorders extend the surveillance cameras of Eden-Olympia into the hinterlands of Nice, Cannes, and La Bocca. The video cameras act as visible reminders that these therapy groups are intrinsic to the environment of Eden-Olympia, enabling its methods of surveillance and control to reach out into neighboring spaces. When parties of Eden-Olympia executives film themselves rampaging through local slums, raping, stealing, and instigating brutal attacks on impoverished communities, the forays represent an extension of Eden-Olympia's systems of control. In Eden-Olympia, instances of "controlled psychopathy" represent a fantasy of total identification and regulation, an effort to transform the most profound instances of antisocial behavior into a means of regulating citizens of the intelligent city. This is implicit in the notion of an intelligent environment, whose infrastructure is designed to identify and neutralize any potential threats to its continuing efficiency and functionality.

By exploring the use of advanced forms of automated monitoring and identification to transform citizens into nodes within a highly regulated cityspace, *Super-Cannes* and *H(A)PPY* show how intelligent environments are predicated on advanced forms of identification and identity formation. These representations of intelligent environments are based on complex processes of monitoring and identification that are themselves subject to struggles for power and ownership. Both novels speculate about the extreme ramifications of an increasingly integrated relationship between citizens and invisible infrastructure, using imagery of pristine environments to represent seeming spaces of perfection, with citizens willingly exchanging anonymity and a right to remain unmonitored for the promise of physical well-being and security. Whereas *H(A)PPY* explores the ramifications of advanced automation and a sensor-based environment via depiction of a System whose inflexibility means it cannot incorporate glitches or breakdowns, *Super-Cannes* offers a sophisticated critique of the principle of incorporating violence to inoculate against threats. Ultimately, both novels offer a scathing warning about citizens' acceptance of complicity with surveillance assemblages in exchange for promises of security, efficiency, and physical well-being.

Notes

1. Anthony M. Townsend, *Smart Cities: Big Data, Civic Hackers, and the Quest for a New Utopia* (New York: W. W. Norton, 2013), 15.

2. Antoine Picon, *Smart Cities: A Spatialised Intelligence* (Chichester, UK: John Wiley and Sons, 2015), 13.

3. Matthew Jewell, "Contesting the Decision: Living in (and Living with) the Smart City," *International Review of Law, Computers and Technology* 32, nos. 2–3 (2 September 2018): 210, doi:10.1080/13600869.2018.1457000.

4. Marcus Foth, "Early Experiments Show a Smart City Plan Should Start with People First," *The Conversation*, 1 June 2016, http://theconversation.com/early-experiments-show-a-smart-city-plan-should-start-with-people-first-60174.

5. J. G. Ballard, *Super-Cannes* (London: Fourth Estate, 2011), 16.

6. Ibid.

7. Hans Schaffers, Carlo Ratti, and Nicos Komninos, "Smart Applications for Smart Cities: New Approaches to Innovation," *Journal of Theoretical and Applied Electronic Commerce Research* 7, no. 3 (December 2012): ii, doi:10.4067/S0718-18762012000300005.

8. J. G. Ballard and John Sutherland, "Scanning the Empty Road with Binoculars," *Good Book Guide*, September 2000.

9. For more on the city as laboratory, see references to smart city "experiments" in Santander, Bristol, Los Angeles, and Toronto, for example in D. Carboni et al., "Scripting a Smart City: The CityScripts Experiment in Santander," in *2013 27th International Conference on Advanced Information Networking and Applications Workshops* (New York: IEEE, 2013), 1265–70, doi:10.1109/WAINA.2013.85; John Murray Brown, "Bristol to Become Smart City Laboratory," *Financial Times*, 30 October 2014, https://www.ft.com/content/44e49120-6034-11e4-98e6-00144feabdc0; and Katherine Peinhardt, "Google's Urban Experiment in Toronto: A Q&A with Sidewalk Labs' Rit Aggarwala," *Medium*, 5 December 2017, https://medium.com/@PPS_Placemaking/googles-urban-experiment-in-toronto-a-q-a-with-sidewalk-labs-rit-aggarwala-5582ae789d53. The "laboratory" tag is not necessarily used enthusiastically. For example, in a 2018 article, Saskia Naafs drew attention to smart cities' contravention of personal data protection, noting that "people should be notified in advance of data collection and the purpose should be specified—but in Stratumseind, as in many other 'smart cities,' this is not the case." "'Living Laboratories': The Dutch Cities Amassing Data on Oblivious Residents," Cities, *Guardian*, 1 March 2018, https://www.theguardian.com/cities/2018/mar/01/smart-cities-data-privacy-eindhoven-utrecht.

10. Jeannette Baxter, *J. G. Ballard: Contemporary Critical Perspectives* (London: Bloomsbury, 2008), 14.

11. Frida Beckman, "Chronopolitics," *Symplokē* 21, nos. 1–2 (2013): 272, doi:10.5250/symploke.21.1-2.0271.

12. Ibid., 7.

13. Ballard, *Super-Cannes*, 39.

14. Lee Rozelle, "'I Am the Island': Dystopia and Ecocidal Imagination in *Rushing to Paradise*, *Super-Cannes*, and *Concrete Island*," *ISLE: Interdisciplinary Studies in Literature and Environment* 17, no. 1 (1 January 2010): 62, doi:10.1093/isle/isp153.
15. Ballard, *Super-Cannes*, 5, 202.
16. Ibid., 356.
17. Le Corbusier, *Towards a New Architecture*, trans. Frederick Etchells (New York: Dover, 1986), 19.
18. Yevgeny Zamyatin, *We*, trans. Natasha Randall (London: Vintage, 2010), 6, 51.
19. Ibid., 16.
20. Andrzej Gąsiorek, *J. G. Ballard* (Manchester: Manchester University Press, 2005), 20.
21. Ballard, *Super-Cannes*, 154.
22. Ibid., 19.
23. Michael M. Wagner, Andrew W. Moore, and Ron M. Aryel, eds., *Handbook of Biosurveillance* (San Diego: Elsevier Science and Technology, 2011), 3.
24. Ibid., 122.
25. Nicola Barker, *H(A)PPY* (London: William Heinemann, 2017), 2, 27.
26. Ibid., 1.
27. Ibid., 101, 176.
28. Ibid., 19.
29. Ballard, *Super-Cannes*, 111.
30. Sungmee Park, Kyunghee Chung, and Sundaresan Jayaraman, "Chapter 1.1—Wearables: Fundamentals, Advancements, and a Roadmap for the Future," in *Wearable Sensors*, ed. Edward Sazonov and Michael R. Neuman (Oxford, UK: Academic Press, 2014), 5, doi:10.1016/B978-0-12-418662-0.00001-5.
31. Townsend, *Smart Cities*, xi.
32. Barker, *H(A)PPY*, 50.
33. Ibid., 220.
34. Ibid., 216.
35. Ibid.
36. Ibid., 220.
37. Jathan Sadowski and Frank Pasquale, "The Spectrum of Control: A Social Theory of the Smart City," *First Monday* 20, no. 7 (24 June 2015), http://firstmonday.org/ojs/index.php/fm/article/view/5903.
38. Patrick Joyce, *The Rule of Freedom: Liberalism and the Modern City* (London: Verso, 2003), 11.
39. Mark Poster, "Databases as Discourse; or, Electronic Interpellations," in *Computers, Surveillance, and Privacy*, ed. David Lyon and Elia Zureik (Minneapolis: University of Minnesota Press, 1996), 184.
40. Ibid.
41. Barker, *H(A)PPY*, 3.
42. Poster, "Databases as Discourse," 184.
43. Barker, *H(A)PPY*, 87.
44. Tim Jordan, *Cyberpower: The Culture and Politics of Cyberspace and the Internet* (London: Routledge, 2002), 205.
45. Barker, *H(A)PPY*, 90.
46. Michel Foucault, *Discipline and Punish: The Birth of the Prison*, trans. Alan Sheridan, 2nd ed. (New York: Vintage, 1995), 201.
47. Ibid., 206.
48. Clare Birchall, *Shareveillance: The Dangers of Openly Sharing and Covertly Collecting Data*, Forerunners: Ideas First (Minneapolis: University of Minnesota Press, 2017), 5.
49. Ibid.
50. Ballard, *Super-Cannes*, 57.
51. Jon Coaffee, David Murakami Wood, and Peter Rogers, *The Everyday Resilience of the City: How Cities Respond to Terrorism and Disaster* (London: Palgrave Macmillan, 2008), 91.
52. Ballard, *Super-Cannes*, 38.
53. Ibid., 84.
54. Ibid., 19.
55. Jewell, "Contesting the Decision," 210.
56. Martijn de Waal, Michiel de Lange, and Matthijs Bouw, "The Hackable City: Citymaking in a Platform Society," *Architectural Design* 87, no. 1 (1 January 2017): 56, doi:10.1002/ad.2131.
57. Steve Pile, Christopher Brook, and Gerry Mooney, *Unruly Cities? Order/Disorder* (London: Routledge, 1999), 98.
58. Myria Georgiou, "Urban Encounters: Juxtapositions of Difference and the Communicative Interface of Global Cities," *International Communication Gazette* 70, nos. 3–4 (1 June 2008): 224, doi:10.1177/1748048508089949.

59. Barker, *H(A)PPY*, 220.
60. Ibid., 44.
61. Ibid., 218.
62. Ibid., 107.
63. Ibid., 98.
64. Ballard, *Super-Cannes*, 154.
65. Ibid., 135.
66. Barker, *H(A)PPY*, 46.
67. Evgeny Morozov, *To Save Everything, Click Here: Technology, Solutionism, and the Urge to Fix Problems That Don't Exist* (London: Allen Lane, 2013), 204–5.
68. Ballard, *Super-Cannes*, 19.
69. Alondra Nelson, "Introduction: Future Texts," *Social Text* 20, no. 2 (2002): 1.
70. Ballard, *Super-Cannes*, 18.
71. Ibid., 58, 251.
72. Ibid., 250–51, 253–54.
73. Ibid., 264.
74. Ibid., 255.
75. Eric A. Ostrowidzki, "Utopias of the New Right in J. G. Ballard's Fiction," *Space and Culture* 12, no. 1 (February 2009): 6, doi:10.1177/1206331208327745.
76. Ibid., 19.

References

Ballard, J. G. *Super-Cannes*. London: Fourth Estate, 2011.

Ballard, J. G., and John Sutherland. "Scanning the Empty Road with Binoculars." *Good Book Guide*, September 2000.

Barker, Nicola. *H(A)PPY*. London: William Heinemann, 2017.

Baxter, Jeannette. *J. G. Ballard: Contemporary Critical Perspectives*. London: Bloomsbury, 2008.

Beckman, Frida. "Chronopolitics." *Symplokē* 21, nos. 1–2 (2013): 271–89. doi:10.5250/symploke.21.1-2.0271.

Birchall, Clare. *Shareveillance: The Dangers of Openly Sharing and Covertly Collecting Data*. Forerunners: Ideas First. Minneapolis: University of Minnesota Press, 2017.

Brown, John Murray. "Bristol to Become Smart City Laboratory." *Financial Times*, 30 October 2014. https://www.ft.com/content/44e49120-6034-11e4-98e6-00144feabdco.

Carboni, D., A. Pintus, A. Piras, A. Serra, A. Badii, and M. Tiemann. "Scripting a Smart City: The CityScripts Experiment in Santander." In *2013 IEEE 27th International Conference on Advanced Information Networking and Applications Workshops*, 1265–70. New York: IEEE, 2013. doi:10.1109/WAINA.2013.85.

Coaffee, Jon, David Murakami Wood, and Peter Rogers. *The Everyday Resilience of the City: How Cities Respond to Terrorism and Disaster*. London: Palgrave Macmillan, 2008.

Foth, Marcus. "Early Experiments Show a Smart City Plan Should Start with People First." *The Conversation*, 1 June 2016. http://theconversation.com/early-experiments-show-a-smart-city-plan-should-start-with-people-first-60174.

Foucault, Michel. *Discipline and Punish: The Birth of the Prison*. Translated by Alan Sheridan. 2nd ed. New York: Vintage, 1995.

Gąsiorek, Andrzej. *J. G. Ballard*. Manchester: Manchester University Press, 2005.

Georgiou, Myria. "Urban Encounters: Juxtapositions of Difference and the Communicative Interface of Global Cities." *International Communication Gazette* 70, nos. 3–4 (1 June 2008): 223–35. doi:10.1177/1748048508089949.

Jewell, Matthew. "Contesting the Decision: Living in (and Living with) the Smart City." *International Review of Law, Computers and Technology* 32, nos. 2–3 (2 September 2018): 210–29. doi:10.1080/13600869.2018.1457000.

Jordan, Tim. *Cyberpower: The Culture and Politics of Cyberspace and the Internet*. London: Routledge, 2002.

Joyce, Patrick. *The Rule of Freedom: Liberalism and the Modern City*. London: Verso, 2003.

Le Corbusier. *Towards a New Architecture*. Translated by Frederick Etchells. New York: Dover, 1986.

Morozov, Evgeny. *To Save Everything, Click Here: Technology, Solutionism, and the Urge to Fix Problems That Don't Exist*. London: Allen Lane, 2013.

Naafs, Saskia. "'Living Laboratories': The Dutch Cities Amassing Data on Oblivious Residents." Cities. *Guardian*, 1 March 2018. https://www.theguardian.com/cities/2018/mar/01/smart-cities-data-privacy-eindhoven-utrecht.

Nelson, Alondra. "Introduction: Future Texts." *Social Text* 20, no. 2 (2002): 1–15.

Ostrowidzki, Eric A. "Utopias of the New Right in J. G. Ballard's Fiction." *Space and Culture* 12, no. 1 (February 2009): 4–24. doi:10.1177/1206331208327745.

Park, Sungmee, Kyunghee Chung, and Sundaresan Jayaraman. "Chapter 1.1—Wearables: Fundamentals, Advancements, and a Roadmap for the Future." In *Wearable Sensors*, edited by Edward Sazonov and Michael R. Neuman, 1–23. Oxford, UK: Academic Press, 2014. doi:10.1016/B978-0-12-418662-0.00001-5.

Peinhardt, Katherine. "Google's Urban Experiment in Toronto: A Q&A with Sidewalk Labs' Rit Aggarwala." *Medium*, 5 December 2017. https://medium.com/@PPS_Placemaking/googles-urban-experiment-in-toronto-a-q-a-with-sidewalk-labs-rit-aggarwala-5582ae789d53.

Picon, Antoine. *Smart Cities: A Spatialised Intelligence*. Chichester, UK: John Wiley and Sons, 2015.

Pile, Steve, Christopher Brook, and Gerry Mooney. *Unruly Cities? Order/Disorder*. London: Routledge, 1999.

Poster, Mark. "Databases as Discourse; or, Electronic Interpellations." In *Computers, Surveillance, and Privacy*, edited by David Lyon and Elia Zureik, 175–92. Minneapolis: University of Minnesota Press, 1996.

Rozelle, Lee. "'I Am the Island': Dystopia and Ecocidal Imagination in *Rushing to Paradise*, *Super-Cannes*, and *Concrete Island*." *ISLE: Interdisciplinary Studies in Literature and Environment* 17, no. 1 (1 January 2010): 61–71. doi:10.1093/isle/isp153.

Sadowski, Jathan, and Frank Pasquale. "The Spectrum of Control: A Social Theory of the Smart City." *First Monday* 20, no. 7 (24 June 2015). http://firstmonday.org/ojs/index.php/fm/article/view/5903.

Schaffers, Hans, Carlo Ratti, and Nicos Komninos. "Smart Applications for Smart Cities: New Approaches to Innovation." *Journal of Theoretical and Applied Electronic Commerce Research* 7, no. 3 (December 2012): ii–v. doi:10.4067/S0718-18762012000300005.

Townsend, Anthony M. *Smart Cities: Big Data, Civic Hackers, and the Quest for a New Utopia*. New York: W. W. Norton, 2013.

Waal, Martijn de, Michiel de Lange, and Matthijs Bouw. "The Hackable City: Citymaking in a Platform Society." *Architectural Design* 87, no. 1 (1 January 2017): 50–57. doi:10.1002/ad.2131.

Wagner, Michael M., Andrew W. Moore, and Ron M. Aryel, eds. *Handbook of Biosurveillance*. San Diego: Elsevier Science and Technology, 2011.

Zamyatin, Yevgeny. *We*. Translated by Natasha Randall. London: Vintage, 2010.

11.
Jennifer Egan and the Database

Rob Lederer

Toward the end of "Love in the Time of No Time," her 2003 article about the rise of online romance, Jennifer Egan imagines the commercial "long-term vision" of internet dating websites, in which they function as

> a virtual clearinghouse where potential lovers, friends, business associates, audience members and devotees of all forms of culture—invisible to one another in the shadowy cracks of cities around the world—are registered, profiled and findable. An alternate dimension where the randomness and confusion of urban life are at last sorted out.[1]

In Egan's early examination of online dating, such websites respond to our desire to overcome the anonymity of the city, to make the street's atomized crowds legible so that the forging of personal connection, romance, and friendship need not rely on chance encounters. Egan returns to this vision of a technologically mapped social landscape in her novel *A Visit from the Goon Squad* (2010). In a chapter set in the mid-1990s, the character Bix fantasizes about a future in which, through digital technology, "we'll rise up out of our bodies and find each other again in spirit form. We'll meet in that new place, all of us together, and first it'll seem strange, and pretty soon it'll seem strange that you could ever lose someone, or get lost."[2] There is optimism in imagining a future in which search engines and personal profiles help combat loneliness and sustain relationships. But these digital fantasies of a city illuminated by data, where it is impossible to wander astray, are also deeply entwined with the encroachments of digital surveillance. The shadow side of Bix's connected spirits is what Kevin D. Haggerty and Richard V. Ericson have called the "progressive 'disappearance of disappearance'—a process whereby it is increasingly difficult for individuals to maintain their anonymity or to escape the monitoring of social institutions."[3]

In this chapter, I will analyze Egan's representation of digital surveillance in *A Visit from the Goon Squad*, her Pulitzer Prize–winning novel. I argue that the database figures centrally in Egan's portrayal of our digitized future as the site where self-formation and social power collide. David Lyon suggests that as surveillance has been increasingly digitized, the database, the technology by which personal information is stored and analyzed, has become a key substrate in surveillance processes. He writes, "Surveillance today is found in the flows of data within networked databases, but these still relate to organizational practices, power and, of course, the persons to whom those data refer."[4] In *Goon Squad*'s final chapter, the characters have had their information tracked and logged in corporate databases, profiles that they can neither access nor understand.[5] Egan homes in on the restrictive idiom of data, into which the self is translated when it is digitally logged in the database. The form of Egan's novel itself operates as a kind of database, and so, in both the novel's narrative and its structure, Egan construes the database as a new site of control and a source of personal alienation.

In *Goon Squad*, Egan situates the ascent of digital data collection within a longer history of surveillance. In his 1994 book, *The Electronic Eye*, Lyon highlights the ways that data-collection practices have become routine components of everyday life, in such ordinary activities as withdrawing money from a bank machine, driving a car, and checking out a library book.[6] The gathering of personal data has proliferated under the operations of "Big Data," which, as Andrew Guthrie Ferguson writes, functions through "the collection and analysis of large data sets with the goal to reveal hidden patterns or insights," producing information that is both financially lucrative and personally intrusive.[7]

These increasingly invasive data-gathering operations are the lifeblood of a pervasive economic system that Shoshana Zuboff outlines in *The Age of Surveillance Capitalism* (2019). According to Zuboff, crucial to the development of "surveillance capitalism" was Google's discovery, in the early 2000s, of how to make use of the behavioral data it collected not just to fine-tune its services but also to generate predictions about its users' future actions and decisions.[8] This operation was initially employed by Google to target advertisements to individual users based on their online behavior. But Zuboff argues that its lessons about the financial value of personal data and the predictions it can be used to calculate underpin "surveillance capitalism," propelling the commercial race to render data from ever more intimate spaces in our lives: our homes, our bodies, and even our inner selves. Zuboff writes, "Surveillance capital wants more than my body's coordinates in time and space. Now it violates the inner sanctum as machines and their algorithms decide the meaning of my breath and my eyes, my jaw muscles, the hitch in my voice, and the exclamation points that I offered in innocence and hope."[9] The ultimate logic of surveillance capitalism is the perfection of its predictive capabilities to produce "guaranteed outcomes," which undermine our essential freedom to project ourselves into an open, uncertain future.[10]

Egan published *Goon Squad* three years before Edward Snowden's disclosure of the data-collecting operations of the National Security Agency, while the processes of surveillance capitalism were still being formed. Nevertheless, in the novel's treatment of data surveillance, we can see a prelude to Zuboff's critiques about the pervasiveness of data extraction, the mysteriousness of its operations, and its assault on self-determination. Egan is interested not only in the database's capacity to track individual users but also in its false promise that the clear, objective language of data can represent complex, disorderly human subjects. *Goon Squad* consistently warns against subscribing to the ideology of the digital screen and to the belief in the "pure" transparency of its aesthetic and its language, endorsing instead the productivity of gaps, indeterminacy, and silence found in the analogue, material world. *Goon Squad* positions its critique of database surveillance within broader dialectics of gathering and dispersal, of clarity and muddiness, and of coherence and opacity. To elude the logic of database surveillance, it advocates investing the self in material collections, in which personal meaning is rendered not in the form of information but in a private language—usually the language of objects—that is illegible to others.

Panopticism and Beyond

Goon Squad is composed of thirteen stylistically distinct chapters that are arranged out of chronological sequence, hopping backward and forward in time and moving in and out of the lives of a large cast of interconnected characters. Over the duration of its time line, which runs between the 1970s and the near future, *Goon Squad* measures the transformation of surveillance into "dataveillance," from traditional modes of watching, spying, and photographing to the collecting and monitoring of personal data.[11] While in *Goon Squad*'s chronologically earlier chapters the gaze is the primary mode of surveillance and self-scrutiny, in its final chapter, "Pure Language," set in the future, we find the possibilities of visualization altered, the social landscape comprehensively mapped, and the self reconceptualized as a compilation of data. That is to say, *Goon Squad* demonstrates the extension of Foucault's model of the panopticon to what Mark Poster calls the superpanopticon, in which databases, used to gather and store information about members of the public and to exert power over them, shift and expand the scope of surveillance.[12]

Michel Foucault famously derived a theory of surveillance, power, and subjection from Jeremy Bentham's panopticon, his blueprint for a prison that places convicts under the unwavering gaze of a central watchtower into which they cannot see. In the panopticon, prisoners live with the sense of being ceaselessly monitored without the ability to tell when they are actually being observed, such that they begin to internalize authority and exercise it on themselves.[13] For Foucault, in modernity the principles of the panopticon extend beyond the walls of the prison and become a general means of control. *Goon Squad* reflects Foucault's understanding of panoptic subjection in its portrayal of characters whose behavior is conditioned by the gazes of others.[14] In the chapter "Goodbye, My Love," Sasha derives comfort from imagining that her lost father is following her on her travels abroad, "making sure [she] was okay" (233). When she begins university, in "Out of Body," Sasha reveals to her friend Rob that her stepfather "told her he was hiring a detective to make sure she 'toed the line' on her own in New York" (198). The detective, though likely Sasha's invention, serves to guide her behavior away from her darker impulses. She, for instance, goes out for steamed vegetables so he can see her "eating healthy food," and when Rob questions her about the detective, she says, "I want him to see me well again—how I'm still normal, even after everything" (202–3).

While Sasha streamlines her behavior because of a real or imagined observer, Rob, in the same chapter, acts as his own inspector, internalizing the regulatory panoptic gaze and directing it at himself. Rob narrates "Out of Body" in the second person, suggesting that he is telling the story to himself, formally re-creating his division in the chapter between participant in the scene and observer of it. When Rob attempts an apology, he narrates to himself: "You're not completely there—a part of you is a few feet away, or above, thinking, Good, they'll forgive you, they won't desert you, and the question is, which one is really 'you,' the one saying and doing whatever it is, or the one watching?" (197). Here Rob appears as a perfect specimen of panopticism, alienated from his own actions and desires because he is also always judging them. When he reveals a past homosexual encounter to Sasha, he narrates: "It wasn't you in the car with James. You were somewhere else, looking down, thinking, That fag is fooling around with another guy" (200). Rob, torn between personal desire and social prejudice, embodies the self-regulatory dynamics of panopticism, perpetually watching himself from a distance and evaluating his behavior according to internalized codes.

The chapter "Selling the General" dramatizes digital technology's intensification of visual surveillance and its regulatory effects. When the publicist Dolly is hired to reform the image of General B, a murderous dictator, she manipulates the public's sympathy by having him photographed wearing a fuzzy hat and later by arranging for him to be accompanied by the celebrity Kitty Jackson. When B takes Kitty hostage, Dolly emails pictures of the pair "nuzzling" to tabloids, which hours later are "being posted and traded on the Web" (172). This increased exposure ironically results in B's policing by newspaper photographers, who find him where assassins never could. Unable to escape their attentive lenses, B must amend his violent activities and transition his country to democracy. Here the policing gaze is redemptive, but the implication nonetheless is that the watchful eyes of others result in the recalibration of behavior. In Dolly's case, the interpenetration of surveillance by corporate press agencies and the internet exponentially increases its scale and speed—when Dolly emails the photos, they are uploaded and exchanged online "within a couple of hours," the international press begins contacting her "by nightfall" (172), and photographers start canvassing the general after "three or four days" (173).

In *Goon Squad*, digital technology enhances the possibilities of surveillance, augmenting the range of what is visible. A leap forward in time in "Safari" reveals the future invention of a surveillance tool by the grandson of a Samburu warrior:

> Joe will go to college at Columbia and study engineering, becoming an expert in visual robotic technology that detects the slightest hint of irregular movement (the legacy of a childhood spent scanning the grass for lions). He'll marry an American

named Lulu and remain in New York, where he'll invent a scanning device that becomes standard issue for crowd security. (65)

A version of Joe's invention appears in "Pure Language," the final, futuristic chapter, in which "visual scanning devices affixed to cornices, lampposts, and trees" monitor the crowd at Scotty's concert (339). The technological swelling of the visual field is similarly apparent when Alex uses his smartphone-like "T handset" to observe his wife, Rebecca, and their daughter at the concert. His T is able to locate his wife's handset within the crowd, and he then spots her using its zoom function—we learn that "without the zoom, he couldn't even see them" (345). The T permits Alex not just to find his wife in the throng but also to observe her voyeuristically as she dances and receives a message from him.

Although these moments demonstrate new capacities for visualization in the digital age, "Pure Language" suggests that new technologies do not present a straightforward enlargement of the field of vision. Zygmunt Bauman argues that the panopticon model, which functions according to a spatialized relationship between those with and those without power, is altered by new technologies that enable the instantaneous transfer of information and eliminate the importance of space, thus allowing control to be asserted from anywhere. Whereas the panopticon is, in Bauman's words, "a model of mutual engagement and confrontation" between the observer and the observed, new technologies permit those in positions of power to remain remote, inaccessible, and invisible.[15] Lyon therefore notes in his analysis of Bauman's work that digital surveillance produces a disparity in transparency: "As the details of our daily lives become more transparent to the organizations surveilling us, their own activities become less and less easy to discern."[16] Zuboff argues that these asymmetries in visibility are also "asymmetries of knowledge and power," which give a small population, acting beyond our direct awareness, unique access to the public's personal data and the knowledge derived from it.[17]

In her journalism, Egan discusses the new forms of visibility and invisibility that characterize online life. For Egan, the ability to compose oneself anew on internet forums, to invent a name and an identity or several names and several identities, troubles the possibility of real intimacy. In "Lonely Gay Teen Seeking Same," her 2000 article about the online lives of gay adolescents, Egan writes that "while the Internet provides a safe haven for countless gay teenagers who don't dare confide their sexual orientations to the people around them, it is also a very easy place to get burned. It's not just that people disappear—it's that in the end, you're never really sure who they were in the first place. And they don't really know you."[18] By shading users from view, the internet provides them simultaneously with a place to reveal their secrets and hide themselves behind a fake identity, producing an environment of uncertainty even as it operates as a space of personal revelation. In this article and in her article about online dating, Egan dissects the tensions between fact and fiction and between visibility and obscurity that structure online relationships. Our screens, Egan's works show, can function both as cloaks, which veil users from each other, and as microscopes, which open up users' lives to the scrutiny of surveillance organizations.

Goon Squad illustrates Bauman's idea that digital technology illuminates the world unevenly—surveillance practices remain opaque, while the institutions using them gain greater access to their subjects—in its representation of the "blind team" advertising scheme, which, as Katherine D. Johnston notes, "draws attention to the hierarchies of communication, visibility, and capital that are more often than not concealed."[19] Music executive Bennie hires Alex to compile a list of people with social influence whom he will pay to advertise Scotty's concert by intimating genuine enthusiasm for it online. This group of inconspicuous promoters are, in the language of the chapter, the "blind team"; individually they are "parrots," having sold their voices to Bennie to advertise the upcoming event. Alex provides the list of possible parrots to Lulu, who contacts them individually, meaning that actors are neither aware of Alex's place on top of the pyramid nor able to identify any of the other parrots. It is a scheme that is enabled by digital technology, and it is, as Alex notes, structured to make the parrots feel less guilty about selling their voices and opinions, to "reduce the shame and guilt of parrothood by assembling a team that doesn't know it's a team—or that it has a captain" (326). The practice of purchasing opinions is prevalent in this future world; the text alludes to the "Bloggescandals," in which, it hints, political commentators sold their advocacy, resulting in a widespread "suspicion that people's opinions weren't really their own" (322). Here we see the atmosphere of distrust, which in Egan's journalism pervaded online relationships, bleed into the real world.

The T device facilitates the construction of the blind team not only technologically but also morally; it permits people to communicate while veiled by the screen, and this isolation encourages them to undermine their ethical beliefs more easily by hiding misdemeanors in the digital sphere. Alex is comforted that, even if Scotty is a disappointing performer, no one will know that he was responsible for leading them to the concert: he thinks, in the language of the T devices, *"no 1 nOs abt me. Im invysbl"* (338). Alex's invisibility, as well as that of Zeus and the other parrots, reveals the unequal availability of invisibility in the digital age, with the marketers operating out of view of the public, creating a milieu of distrust whereby anyone's opinion might be as genuine as an advertisement. By contrast, Egan emphasizes the public's increased visibility to institutions when Alex links his own disaffection to digital surveillance. He asks himself whether his willingness to compromise his previously high morals by organizing the blind team comes from his inability to

> forget that every byte of information he'd posted online (favorite color, vegetable, sexual position) was stored in the databases of multinationals who swore they would never, ever use it—that he was *owned*, in other words, having sold himself unthinkingly at the very point in his life when he'd felt most subversive? (324)

Here, Alex articulates his enforced blindness toward and uncertainty about the purposes to which his data is being put. Alex experiences his loss of information to the database as a loss of self, such that he can consciously betray his former ideals.

Mark Poster offers an account of the database that illuminates Egan's vision of digital surveillance and signification. Poster argues that databases reconstitute individuals according to an informational architecture of "rigidly defined categories or fields," comprised of limited space and filled out in a restricted data language, which reduces their subjects' complexity.[20] For Poster, the violence of the database comes from its "non-ambiguous grammatical structure" and the assumption that this abbreviated informational language can capture the self unequivocally.[21] Alex's précis of his database estrangement ("favorite color, vegetable, sexual position") parodies both the invasiveness of surveillance and the limited parameters by which the database outlines its subjects. The process of being databased in *Goon Squad* is fundamentally alienating. When Alex agrees to organize the blind team, in opposition to his ethical values, it is as though he is reinventing himself in response to his loss of identity, to its capture in corporate databases.

We can find an earlier version of Alex's alienation by the database in Egan's novel *Look at Me*, in which the model Charlotte, who looks drastically different following a car crash, is given the chance to diarize her life for the public on the website Extraordinary People. Tellingly, Thomas, the company's founder, says of the site that "it's not a magazine—it's a database."[22] The form of Charlotte's own page, called a PersonalSpace, structures several of the categories through which she conceptualizes and displays her life—categories that include "Childhood Memories. Dreams. Diary Entries . . . Future Plans/Fantasies. Regrets/Missed Opportunities."[23] Just after her meeting with Thomas, Charlotte begins thinking of her past using these key terms, or, as she states, "formatting my thoughts to Thomas Keene's specifications and calculating their price."[24] In this moment, we see the power of the database's logic, as Charlotte reforms her memory and her ways of making sense of the past according to its structure in a bid to maximize her financial profits.

The process of displaying her life online precipitates a stark division in Charlotte between herself as an actor in the world and as a self-archivist, and she ends the novel by selling her image and her past to Thomas for the distinctly panoptic reason that "life can't be sustained under the pressure of so many eyes."[25] Like Charlotte, who ends the narrative severed from her previous self, Alex feels compelled to remodel his behavior because his data has been captured in official databases. While Charlotte knowingly published her life for the public, in *Goon Squad* we see Alex grapple with the repercussions of a more clandestine process of data collection. Alex is no longer naïve about digital surveillance, but his information has already been taken and he is powerless both to reclaim it and to know precisely how it is being used.

In an earlier chapter in *Goon Squad*, Egan provides a commentary on the system of database surveillance in the monitoring apparatus that Jules Jones imagines for Central Park. In Jules's dystopian vision, encoded checkpoints will measure from a bank of records the trustworthiness of each person in the park by calculating a numbered ranking from a list of increasingly ridiculous categories that include:

> marriage or lack thereof, children or lack thereof, professional success or lack thereof, healthy bank account or lack thereof, contact with childhood friends or lack thereof, ability to sleep peacefully at night or lack thereof, fulfillment of sprawling, loopy youthful ambitions or lack thereof, ability to fight off bouts of terror and despair or lack thereof. (189)

Jules's forecast animates Poster's assertion that databases delimit their subjects through a symbolic system that "contains no ambiguity," numerical grades being a primary example.[26] Such rankings, Poster posits, show the database to be shaped by the political associations of its owners, who assign the reductive number grades and thereby constitute subjects according to the owners' ideologies, in this case their ideas about what amounts to a successful life. This allocation of rankings creates real-world effects: Jules's system, complemented by radar screens and security guards, regulates the movement of parkgoers by ensuring that the nonfamous do not bother the celebrities with higher rankings. Thus, we find in Jules's system the realization of Poster's idea that the database sees "the constitution of an additional self, one that may be acted upon to the detriment of the 'real' self without that 'real' self ever being aware of what is happening."[27]

The "non-ambiguous" grammar of data is, in "Pure Language," not confined to the database but has come to act as an aesthetic ideal for Lulu and her generation and to inform their thinking about communication over their T devices. Walter Benjamin defines information as "lay[ing] claim to prompt verifiability," suggesting that it assumes the appearance of being "understandable in itself."[28] It is just this quality of clarity that Lulu appreciates in the language of the T handset. When Lulu casually insults Alex while debating the ethics of the blind team, she reflects on the failures of spoken conversations: "There are so many ways to go wrong. . . . All we've got are metaphors, and they're never exactly right. You can't ever just *Say. The. Thing*" (328, emphasis in original). She continues her interaction with Alex, and they exchange the list of parrots, over their T devices, because for Lulu this mode of communication is "pure—no philosophy, no metaphors, no judgments" (329). Lulu's veneration of the language of the T device for its directness connects to the novel's more general commentary on digital aesthetics. Music-industry tycoon Bennie laments the loss of "muddiness" in digital recordings, "the sense of actual musicians playing actual instruments in an actual room" (23). He complains: "Too clear, too clean. The problem was precision, perfection; the problem was *digitization*, which sucked the life out of everything that got smeared through its microscopic mesh. Film, photography, music: dead. *An aesthetic holocaust!*" (24, emphasis in original). The discrepancy between the complication of the real world and the unspoiled sheen of the digital is similarly expressed by Jocelyn, who sees in Lou's "new, flat and long" television "a nervous sharpness that makes the room and even us look smudged" (91). Here, Bennie and Jocelyn echo an observation made by Jacques Derrida, who notes a shift in the palpability of deletions from writing to word processing. While in traditional writing erasures and insertions take physical and mental shape, leaving "a sort of scar on the paper or a visible image in the memory," on the computer screen "everything negative is drowned, deleted; it evaporates immediately, sometimes from one instant to the next."[29]

Lulu and her generation are searching for a language that is, like the digital screen and the music of Bennie's digital "aesthetic holocaust," without blemishes, without signs that any meaning has been hidden or suppressed. This pursuit is evident in their diction, which is wiped of profanities, and on their bodies, which are "'clean': no piercings, tattoos or scarifications" (325).[30] Lulu and her peers likewise disavow the terms "up front" and "out in the open" because in assuming that the motivations and beliefs that underlie speech can be made entirely transparent, such expressions "impl[y] the existence of an ethically perfect state" (327). Lulu instead rejects the hidden and the undisclosed in discourse, and so she has no ethical problem with buying opinions for the blind team; she contends that "if I believe, I believe. Who are you to judge my reasons?" (327). The novel thus links the unmarked aesthetics of digital technology to the denigration of an ethics based on notions of depth and interiority. Note, for instance, in Lulu's formula for the purity of the T language her conjunction of its analytic and aesthetic neutrality ("no philosophy, no metaphors") with amorality ("no judgments").

Like the social terrain of "Pure Language," whose blank spaces seem to have been mapped by database surveillance, Lulu's digitized generation seeks a symbolic system without indeterminacy, a transparent language in which all significance is immediately visible. This is the "pure language" of the chapter's title, and it is part of a more general sense of what we could call, borrowing a term from Avery Gordon, the "hypervisibility" of digital technology—its aesthetic, its language, and its surveillance—in *Goon Squad*. Gordon writes, "In a culture seemingly ruled by technologies of hypervisibility, we are led to believe not only that everything can be seen, but also that everything is available and accessible for our consumption."[31] The myth of the hypervisibility of data language is precisely the problem for Poster, who seeks to demystify the belief that the progression to electronic language will solve the problems of previous forms of communication. Poster argues instead that electronic database language "incurs a tremendous *loss* of data, or better, imposes a strong reading on it," thereby increasing errors and ambiguities.[32]

Lulu buys into the idea that the clarity of data language has solved the problems with spoken conversation, with the evasiveness of metaphor, but in "Pure Language" we see signs of faults in this system of communication.

Alex, for instance, critiques Lulu's partiality for the simplicity of this language when he notes "how easily baby talk fitted itself into the crawl space of a T" (335). Despite the assumption that data illuminates wholly and quantifies accurately, there remain meaningful flickers of the indistinct and unquantifiable. Alex receives a message from Lulu that he reads as "nice," an uncharacteristically sarcastic reply to his lamentations of the dwindling supply of air and light in his apartment—until he understands that, although "*nyc*" can be read as "nice," what Lulu means is "New York City" (335).[33] We find a similar volatility in the numerical language of Jules's Central Park security system. Jules asks that his own infamy for assaulting Kitty Jackson be treated akin to other forms of fame, such that he will be rewarded with a high ranking and additional privilege (189). Jules's surveillance model, based on ostensibly objective indices, thus contains its own slippery indeterminacy, whereby various types of notoriety collapse into each other.

While Lulu's generation reaches out for a language that—like the surveilled space they occupy—is resistant to the hidden or the undisclosed, these examples insist on an undertow to data signification that transcends its surface simplicity. What is condemned is not just the reductive discourse of data but also the belief that this sparse, direct language could achieve comprehensive transparency. Data, rather, is seen as a language that occludes its representational frailties by assuming the capacity to capture truth simplistically, to illuminate perfectly the self and the world around it. The chapter's critique of the database for reducing subjects while naturalizing a belief in the transparency of its idiom is part of a larger proposition the novel makes about the false promises of digital aesthetics and surveillance to reveal the world entirely. The purity of the digital screen so revered by Lulu functions by elision, hiding its representational failures while claiming pellucidity. The form of the novel, as I'll discuss in the next section, functions as a reaction to the "purity" of the digital screen by inserting holes and gaps—or, in Bennie's language, "muddiness"—into its organization. At the same time, it is structured according to a database form that produces its own aesthetic of surveillance.

Digital Aesthetics and the Database

Goon Squad's formal discontinuity casts it against the unnaturally pristine, unmarked quality that it associates with digital technology. The novel's structure is episodic, its chapters formed of moments from the lives of its characters—for example, Lou's trip to Africa and Dolly's handling of the general. In between the chapters we find temporal and narrative fissures. Egan has indicated in interviews that three rules governed *Goon Squad*'s composition: (1) each chapter should take a different protagonist, (2) each should employ a different technical or narrative style, and (3) each could be read in isolation from the collection.³⁴ The novel begins with Sasha, who steals a wallet while on a date with Alex, during which we hear in passing about her ex-boss, Bennie. He is the central character of the second chapter, which goes back in time to a moment when Sasha is still working as his assistant. As the text continues to move backward and forward through time, we encounter narrative styles that are conventional (first person and third person) and experimental (postmodern celebrity profile and PowerPoint presentation).

Egan's compositional rules, which ensure stark separations between chapters caused by disjunctions in time, point of view, tone, and style, provide moments of pause in the text. James Zappen suggests that, like the other silences in the novel, the gaps between the chapters supply readers with space to contemplate the characters' decisions in order to connect the various parts of their narratives.³⁵ Thus, the gaps that Egan inserts into the novel's structure materialize Wolfgang Iser's reader-response theory—his suggestion that it is a literary text's indeterminacy that invites the reader to engage with the work and synthesize it into a coherent whole. For Iser, moments of indeterminacy open up "gaps" in the text, which "give the reader a chance to build his own bridges, relating the different aspects of the object which have thus far been revealed to him."³⁶ Such gaps, Iser writes, can be produced by "the abrupt introduction of new characters or even new threads of the plot, so that the question arises as to the connections between the story revealed so far and the new, unforeseen situations."³⁷ *Goon Squad*'s chapters function in just this way by taking on different protagonists and temporal and physical settings, and asking that the reader make sense of the various interconnections that bind the novel together, however loosely.

In addition to materializing Iser's theory of literary indeterminacy, *Goon Squad*, through the insertion of deep breaks between its chapters, is structured according to the aesthetic of the digital database. In its final two chapters, *Goon Squad* points to the demise of traditional narrative in the digital era. In "Pure Language," Rebecca's academic research indicates that the terms "story" and "change" have been "shucked of their meanings" (331). Lulu, likewise, suggests that the mechanical terms "connect" and "transmit" fail to describe how information travels; "reach," or social influence, she continues, "isn't describable in terms of cause and effect anymore: it's simultaneous" (324–25). The implications for this change are explored in the structure of the preceding chapter, "Great Rock and Roll Pauses," presented in the form of a PowerPoint presentation, which, Egan has remarked, "breaks down a narrative into a sequence of moments that basically hang in the air, and then give up their place to the next moment."[38] Mimicking the text's disjointed structure, Alison's PowerPoint journal necessarily inserts spaces within and between the slides, gaps that interrupt the appearance of a flowing story.[39] The presentation thus exposes the process of literary meaning-making in the rest of *Goon Squad*, requiring that the reader produce a sense of unity out of a set of fractured, discontinuous parts.

Whereas Alison's PowerPoint journal does maintain a stable system of arrangement, *Goon Squad*'s chapters can be reordered, and therefore the novel takes on the basic aesthetic properties of the database. Lev Manovich defines the database as a nonsequential collection that eschews the thematic or formal development of a story, casting it as the key "symbolic form" of the computer era and the enemy of traditional narrative. He writes, "As a cultural form, the database represents the world as a list of items, and it refuses to order this list. In contrast, a narrative creates a cause-and-effect trajectory of seemingly unordered items (events)."[40] In the database, we find an aesthetic mode without formal arrangement, where the sections can be read in any order. Kristin Veel clarifies the temporality of the database in language that connects it to Rebecca's and Lulu's evaluations of digitization, in which "story" has become meaningless, and cause and effect collapse into "simultaneity." Veel writes that "in the database we find that a narrative conception of time as a sequence of causally connected events is replaced by a notion that everything is potentially present at the same time—linearity is replaced by simultaneity."[41] While each of *Goon Squad*'s chapters can be read as an isolated narrative, when appended together they appear as a database whose sections are subject to recombination.

With each chapter, we find the social universe of the novel expanding, introducing new characters and new potential stories. Seen in this way, the novel does not move toward resolution but rather accumulates more and more information, more plot. Manovich suggests that, unlike narratives, which are formally structured to give a sense of development, databases are "collections of individual items, with every item possessing the same significance as any other."⁴² As such, Norman M. Klein, an author and theorist of database novels, writes that data "proceed by insinuation, by involution—toward a *beginning*, toward an *aporia*."⁴³ *Goon Squad* concludes with such a beginning: Bennie and Alex search the night for Sasha but find instead "another girl, young and new to the city, fiddling with her keys" (349). The novel, this final scene indicates, could continue indefinitely, introducing another yet unexplored character. Egan has even suggested that her short story "Black Box," which follows Lulu on an international spying mission and is composed as a series of tweets, could be considered a fourteenth chapter of the book.⁴⁴

Although its numbered chapters are bound in order, I suggest that the database quality of the novel, which would allow readers to reorder its sections, informs any reading of the text, even in its material form. Wolfgang Funk, for instance, suggests that in requiring readers to tease out the various narrative and chronological connections between the chapters, the novel "present[s] human existence as a 'heap of broken stories' without an ordering, authoritative instance to guide the reader through the incongruity of this mortal coil."⁴⁵ The ability to shuffle *Goon Squad*'s chapters, which remains a latent possibility in the physical text, finds expression in the book's iPad application. Released by Egan's publisher, Constable and Robinson, this version allows the reader, after reading through the novel once, to arrange the chapters chronologically or to randomize their order.⁴⁶ The *Goon Squad* app thus makes available the database quality of the novel's structure, subverting the material text's organization and rendering it one of many possible versions.

Goon Squad's material text does not explicitly invite readers to shuffle its chapters, but the possibilities of recombination continue to haunt it, and this implicit database form in turn deflates the appearance of agency among the characters. Cathleen Schine suggests that "because the novel looks both forward and backward in time, and because the facts of its characters' lives are doled out so unexpectedly, so fully out of the obvious sequence of events, the reader acquires an omniscience that is almost godlike, but is nevertheless shadowed with mystery."⁴⁷ The reader's mysterious omniscience, their awareness "that Egan's characters have no choices left," is produced by what I have identified as its database structure, which permits episodes and events to arrive seemingly in any order.⁴⁸ Similarly, Pankaj Mishra writes of *Goon Squad*'s structure that it "leaves us with a disturbing sense of their (and our) state of unfreedom: it shows us the full arc of their choiceless lives."⁴⁹ Zuboff has suggested that the data harvesting and behavioral forecasting of "surveillance capitalism" threatens our "right to the future tense," to imagine our place in a future that has yet to be determined.[50] The novel's database structure, even if it is not fully enabled in the codex version, echoes this vision in so thoroughly disabling temporal chronology, producing a text in which all events have already occurred, with characters stripped of the semblance of free will.

Data and Objects

Clare Birchall has advocated rebelling against and breaking down the system of data-sharing and surveillance, or what she has termed "shareveillance," by seeking out "forms of illegibility: a reimagined opacity."[51] While Birchall lists several technological innovations that disrupt the conventions of data collection, *Goon Squad* presents, as a resistance to database surveillance, the strategy of defining the self in material collections. Jean Baudrillard understands collectors as necessarily isolated creatures, because, he suggests, their objects form a personal language that "has lost any general validity."[52] While for Baudrillard there is something pitiable about the collector's formation of "a discourse addressed to himself," *Goon Squad* locates in the material collection's incomprehensibility to outsiders a corrective to the assumed transparency of database language.[53]

Whereas Alex feels owned by corporations, who have databased his details in easily readable language, material and written collections provide a mechanism for displaying the self that vexes interpretation. The novel's first chapter deals with Sasha's kleptomania, her assembly of stolen objects that contains "the raw, warped core of her life" (16). During their date, she anxiously observes Alex examine her archive, worried that he can decipher its biographical secrets, and she pulls him to the carpet to have sex with him. But as she previously noted, the collection is "illegible yet clearly not random" (15). Alex's stare simply allows him to locate a packet of bath salts, to address the collection as a mess of useful objects. Sasha's arousal seemingly stems from showcasing her biography with the knowledge that it cannot be parsed. She is "drawn to the sight of him taking everything in," but though Alex sees everything, he understands little about it (15). The objects are, however, evocative for Sasha. When Alex uses the bath salts, their aroma reminds Sasha of her former friend Lizzie, their previous owner.

In the next chapter, Sasha misinterprets the meaning of Bennie's collection with similar results. Throughout the chapter, Bennie compiles a list of personal events fraught with shame, denoted by key words: "Kissing Mother Superior, incompetent, hairball, poppy seeds, on the can" (39). Though its language is clipped, it is not the "non-ambiguous" idiom of data described by Poster, meant to present itself plainly to others. Instead, like Sasha's objects, it marks out a space of personal opacity. When Sasha reads the list, she mistakes it for a compilation of potential song titles, an error that motivates Bennie to reconsider his past: "Now they sounded like titles to him, too. He felt peaceful, cleansed" (39). Sasha's "scratchy voice" has "neutralized" the past, and the act of misunderstanding again erupts in a kind of intimacy, a feeling that Benny identifies as love for Sasha (39).

While these two chapters announce an emotional, even erotic, joy in displaying the self in a language illegible to outsiders, "Great Rock and Roll Pauses" locates in the impermanence of objects another revolt against the durability and inaccessibility of the digital record. In this desert future, Sasha makes sculptures out of the flotsam of her family's everyday life that, she says, "tell the whole story if you really look" (273). These archives are not meant to be preserved but instead are designed to degrade and disappear into the desert environment. When Alison and her father return from their nighttime walk, they see Sasha's sculptures "fading into the dust" (294). This is "part of the process," writes Alison, quoting her mother (250). Sasha's collages elude Alex's anxieties about the database through their material ephemerality. In contrast to data's reproducibility, the collages are meant to self-destruct, and whereas Alex cannot be certain how his information is being used, Sasha's artworks ensure that her biography cannot be preserved indefinitely.

By using objects and their evocative personal meanings as the stuff of identity, Sasha has refused to portray herself in a form that might be hacked. Taken together, Sasha's two archives find in the corruptibility of the material collection and the impenetrability of its signification to others a strategy for dodging the mechanisms of database surveillance. Whereas the database's definite language claims to unlock the self, the material collection hides it in an inscrutable and personal idiom that resists interpretation. The personal information contained in the database is long-lasting and reproducible, and the ways it is manipulated are veiled, unlike the information in the material collection, which ultimately erases its own meaning as it decomposes. In contrast to the surveillance enabled by digital technology and the assumed purity of its language, Egan promotes reengaging with the material world and the "muddiness" of its aesthetic. The material archive, she suggests, skirts the assumed readability and durability of the database, providing a form of subjective opacity outside the system of data collection, a place that cannot be spied on and whose meaning cannot be stored and interpreted.

Notes

1. Jennifer Egan, "Love in the Time of No Time," *New York Times Magazine*, 23 November 2003, https://www.nytimes.com/2003/11/23/magazine/love-in-the-time-of-no-time.html.

2. Jennifer Egan, *A Visit from the Goon Squad* (London: Corsair, 2010), 209. All further references to this text are given parenthetically in the chapter.

3. Kevin D. Haggerty and Richard V. Ericson, "The Surveillant Assemblage," *British Journal of Sociology* 51, no. 4 (December 2000): 619.

4. David Lyon, *Surveillance Studies: An Overview* (Cambridge, UK: Polity, 2007), 4.

5. Katherine D. Johnston similarly interprets Egan's novel as a commentary on surveillance. Johnston describes *Goon Squad* as a work of "*profile-fiction*" that reveals and critiques the gendered, racialized, and colonial hierarchies that are perpetuated by systems of surveillance and data profiles. "Metadata, Metafiction, and the Stakes of Surveillance in Jennifer Egan's *A Visit from the Goon Squad*," *American Literature* 89, no. 1 (March 2017): 156 (emphasis in original).

6. David Lyon, *The Electronic Eye: The Rise of Surveillance Society* (Cambridge, UK: Polity, 1994), 4.

7. Andrew Guthrie Ferguson, *The Rise of Big Data Policing: Surveillance, Race, and the Future of Law Enforcement* (New York: New York University Press, 2017), 8.

8. Shoshana Zuboff, *The Age of Surveillance Capitalism: The Fight for a Human Future at the New Frontier of Power* (London: Profile, 2019), 78.

9. Ibid., 290.

10. Ibid., 328–31.

11. Roger A. Clarke explains the concept of dataveillance in "Information Technology and Dataveillance," *Communications of the ACM* 31, no. 5 (May 1988): 499.

12. Mark Poster, *The Mode of Information: Poststructuralism and Social Context* (Cambridge, UK: Polity, 1990), 93.

13. Michel Foucault, *Discipline and Punish: The Birth of the Prison*, trans. Alan Sheridan (London: Penguin, 1991), 202–3.

14. Visual surveillance, as Johnston notes, takes on particular significance during the titular "Safari" of the chronologically earliest chapter. Johnston argues that Egan connects the masculine, imperialistic gaze of the safari to the invasive digital surveillance mechanisms of the novel's final chapter. Johnston, "Metadata," 159.

15. Zygmunt Bauman, *Liquid Modernity* (Cambridge, UK: Polity, 2000), 10.

16. Zygmunt Bauman and David Lyon, *Liquid Surveillance* (Cambridge, UK: Polity, 2013), 12.

17. Zuboff, *Age of Surveillance*, 80–81.

18. Jennifer Egan, "Lonely Gay Teen Seeking Same," *New York Times Magazine*, 10 December 2000, https://www.nytimes.com/2000/12/10/magazine/lonely-gay-teen-seeking-same.html.

19. Johnston, "Metadata," 161.

20. Poster, *Mode of Information*, 96.

21. Ibid.

22. Jennifer Egan, *Look at Me* (London: Corsair, 2001), 245.

23. Ibid.

24. Ibid., 258.

25. Ibid., 514.
26. Poster, *Mode of Information*, 96.
27. Ibid., 97–98.
28. Walter Benjamin, "The Storyteller: Reflections on the Works of Nikolai Leskov," in *Illuminations*, ed. Hannah Arendt, trans. Harry Zorn (1968; New York: Schocken, 2007), 89.
29. Jacques Derrida, *Paper Machine*, trans. Rachel Bowlby (Stanford: Stanford University Press, 2005), 24.
30. Martin Moling likewise notes that Egan unifies the aesthetics of Lulu's generation, the language of the T, and digitized rock music around a common definition of *purity* as "clear, clean, full of precision, and completely inanimate." "'No Future': Time, Punk Rock and Jennifer Egan's *A Visit from the Goon Squad*," *Arizona Quarterly* 72, no. 1 (Spring 2016): 66.
31. Avery Gordon, *Ghostly Matters: Haunting and the Sociological Imagination* (Minneapolis: University of Minnesota Press, 2008), 16.
32. Poster, *Mode of Information*, 94.
33. Moling suggests that this moment of uncertainty, in which the aesthetic "purity" of the T language is undercut by ambiguity and equivocality, reveals that even this condensed technological discourse is capable of supporting poetic expression. "No Future," 66–68.
34. Jennifer Egan, "National Book Award Author Events," interview by Deborah Treisman, podcast audio, 31 March 2011, https://soundcloud.com/nationalbook/jennifer-egan-and-deborah.
35. James P. Zappen, "Affective Identification in Jennifer Egan's *A Visit from the Goon Squad*," *LIT: Literature Interpretation Theory* 27, no. 4 (2016): 300.
36. Wolfgang Iser, *Prospecting: From Reader Response to Literary Anthropology* (Baltimore: Johns Hopkins University Press, 1989), 9.
37. Ibid., 11.
38. Doug Kim, "10 Tips for Great Storytelling from a PowerPoint Novelist," *Microsoft Office Show* (blog), Microsoft Office, 18 August 2010, https://web.archive.org/web/20100821101706/http://blogs.office.com/b/office-show/archive/2010/08/18/10-tips-for-great-storytelling-from-a-powerpoint-novelist.aspx.
39. Egan has called the PowerPoint a "microcosm" of the novel's more general formal concerns. See Egan, "National Book Award."
40. Lev Manovich, *The Language of New Media* (Cambridge, MA: MIT Press, 2001), 225.
41. Kristin Veel, *Narrative Negotiations: Information Structures in Literary Fiction* (Göttingen, DE: Vandenhoeck and Ruprecht, 2009), 91.
42. Manovich, *Language of New Media*, 218.
43. Norman M. Klein, "Waiting for the World to Explode: How Data Convert into a Novel," in *Database Aesthetics: Art in the Age of Information Overflow*, ed. Victoria Vesna (Minneapolis: University of Minnesota Press, 2007), 91.
44. Jennifer Egan, "This Week in Fiction: Jennifer Egan," interview by Deborah Treisman, *Page-Turner* (blog), *New Yorker*, 28 May 2012, http://www.newyorker.com/online/blogs/books/2012/05/this-week-in-fiction-jennifer-egan.html.
45. Wolfgang Funk, "Found Objects: Narrative (as) Reconstruction in Jennifer Egan's *A Visit from the Goon Squad*," in *The Aesthetics of Authenticity: Medial Constructions of the Real*, ed. Wolfgang Funk, Florian Gross, and Irmtraud Huber (Bielefeld, DE: Transcript, 2012), 49–50.
46. Alastair Horne, "A Visit from the Goon Squad," *Bookseller*, 9 May 2011, https://www.thebookseller.com/futurebook/visit-goon-squad.
47. Cathleen Schine, "Cruel and Benevolent," *New York Review of Books*, 11 November 2010, 25.
48. Ibid.
49. Pankaj Mishra, "Modernity's Undoing," *London Review of Books*, 31 March 2011, http://www.lrb.co.uk/v33/n07/pankaj-mishra/modernitys-undoing.
50. Zuboff, *Age of Surveillance*, 328–31.

51. Clare Birchall, *Shareveillance: The Dangers of Openly Sharing and Covertly Collecting Data* (Minneapolis: University of Minnesota Press, 2018), chap. 6, Kindle.

52. Jean Baudrillard, *The System of Objects*, trans. James Benedict (1968; London: Verso, 2005), 114.

53. Ibid.

References

Baudrillard, Jean. *The System of Objects*. Translated by James Benedict. London: Verso, 2005. Originally published 1968.

Bauman, Zygmunt. *Liquid Modernity*. Cambridge, UK: Polity, 2000.

Bauman, Zygmunt, and David Lyon. *Liquid Surveillance*. Cambridge, UK: Polity, 2013.

Benjamin, Walter. "The Storyteller: Reflections on the Works of Nikolai Leskov." In *Illuminations*, edited by Hannah Arendt, translated by Harry Zorn, 83–109. New York: Schocken, 2007. Originally published 1968.

Birchall, Clare. *Shareveillance: The Dangers of Openly Sharing and Covertly Collecting Data*. Minneapolis: University of Minnesota Press, 2018. Kindle.

Clarke, Roger A. "Information Technology and Dataveillance." *Communications of the ACM* 31, no. 5 (May 1988): 498–512.

Derrida, Jacques. *Paper Machine*. Translated by Rachel Bowlby. Stanford: Stanford University Press, 2005.

Egan, Jennifer. "Lonely Gay Teen Seeking Same." *New York Times Magazine*, 10 December 2000. https://www.nytimes.com/2000/12/10/magazine/lonely-gay-teen-seeking-same.html.

———. *Look at Me*. London: Corsair, 2001.

———. "Love in the Time of No Time." *New York Times Magazine*, 23 November 2003. https://www.nytimes.com/2003/11/23/magazine/love-in-the-time-of-no-time.html.

———. "National Book Award Author Events." Interview by Deborah Treisman. Podcast audio, 31 March 2011. https://soundcloud.com/nationalbook/jennifer-egan-and-deborah.

———. "This Week in Fiction: Jennifer Egan." Interview by Deborah Treisman. *Page-Turner* (blog), *New Yorker*, 28 May 2012. http://www.newyorker.com/online/blogs/books/2012/05/this-week-in-fiction-jennifer-egan.html.

———. *A Visit from the Goon Squad*. London: Corsair, 2010.

Ferguson, Andrew Guthrie. *The Rise of Big Data Policing: Surveillance, Race, and the Future of Law Enforcement*. New York: New York University Press, 2017.

Foucault, Michel. *Discipline and Punish: The Birth of the Prison*. Translated by Alan Sheridan. London: Penguin, 1991.

Funk, Wolfgang. "Found Objects: Narrative (as) Reconstruction in Jennifer Egan's *A Visit from the Goon Squad*." In *The Aesthetics of Authenticity: Medial Constructions of the Real*, edited by Wolfgang Funk, Florian Gross, and Irmtraud Huber, 41–61. Bielefeld, DE: Transcript, 2012.

Gordon, Avery. *Ghostly Matters: Haunting and the Sociological Imagination*. Minneapolis: University of Minnesota Press, 2008.

Haggerty, Kevin D., and Richard V. Ericson. "The Surveillant Assemblage." *British Journal of Sociology* 51, no. 4 (December 2000): 605–22.

Horne, Alastair. "A Visit from the Goon Squad." *Bookseller*, 9 May 2011. https://www.thebookseller.com/futurebook/visit-goon-squad.

Iser, Wolfgang. *Prospecting: From Reader Response to Literary Anthropology*. Baltimore: Johns Hopkins University Press, 1989.

Johnston, Katherine D. "Metadata, Metafiction, and the Stakes of Surveillance in Jennifer Egan's *A Visit from the Goon Squad*." *American Literature* 89, no. 1 (March 2017): 155–84.

Kim, Doug. "10 Tips for Great Storytelling from a PowerPoint Novelist." *Microsoft Office Show* (blog). Microsoft Office, 18 August 2010. https://web.archive.org/web/20100821101706/http://blogs.office.com/b/office-show/archive/2010/08/18/10-tips-for-great-storytelling-from-a-powerpoint-novelist.aspx.

Klein, Norman M. "Waiting for the World to Explode: How Data Convert into a Novel." In *Database Aesthetics: Art in the Age of Information Overflow*, edited by Victoria Vesna, 86–94. Minneapolis: University of Minnesota Press, 2007.

Lyon, David. *The Electronic Eye: The Rise of Surveillance Society*. Cambridge, UK: Polity, 1994.

———. *Surveillance Studies: An Overview*. Cambridge, UK: Polity, 2007.

Manovich, Lev. *The Language of New Media*. Cambridge, MA: MIT Press, 2001.

Mishra, Pankaj. "Modernity's Undoing." *London Review of Books*, 31 March 2011. http://www.lrb.co.uk/v33/n07/pankaj-mishra/modernitys-undoing.

Moling, Martin. "'No Future': Time, Punk Rock and Jennifer Egan's *A Visit from the Goon Squad*." *Arizona Quarterly* 72, no. 1 (Spring 2016): 51–77.

Poster, Mark. *The Mode of Information: Poststructuralism and Social Context*. Cambridge, UK: Polity, 1990.

Schine, Cathleen. "Cruel and Benevolent." *New York Review of Books*, 11 November 2010, 25–26.

Veel, Kristin. *Narrative Negotiations: Information Structures in Literary Fiction*. Göttingen, DE: Vandenhoeck and Ruprecht, 2009.

Zappen, James P. "Affective Identification in Jennifer Egan's *A Visit from the Goon Squad*." *LIT: Literature Interpretation Theory* 27, no. 4 (2016): 294–309.

Zuboff, Shoshana. *The Age of Surveillance Capitalism: The Fight for a Human Future at the New Frontier of Power*. London: Profile, 2019.

Contributors

Dorothy Butchard is a lecturer in contemporary literature and digital cultures in the Department of English at the University of Birmingham. Her research explores creative responses to digital technologies, with particular focus on changing power structures in the twentieth and twenty-first centuries.

Patricia E. Chu teaches at Framingham State University. She is the author of *Race, Nationalism and the State in British and American Modernism* and other essays on race and gender theory, ethnic literature, the history of science, and modernism.

Rex Ferguson is senior lecturer in English literature at the University of Birmingham, United Kingdom. He specializes in twentieth-century literature, and his work has appeared in international journals including *New Formations*, *Textual Practice*, *Critical Quarterly*, and the *Journal of Modern Literature*. He is the author of *Criminal Law and the Modernist Novel: Experience on Trial* (Cambridge University Press, 2013).

Jonathan Finn is associate professor in the Department of Communication Studies at Wilfrid Laurier University. He is the author of *Beyond the Finish Line: Images, Evidence, and the History of the Photo-Finish* (McGill-Queen's University Press, 2020) and *Capturing the Criminal Image: From Mugshot to Surveillance Society* (University of Minnesota Press, 2009) and the editor of *Visual Communication and Culture: Images in Action* (Oxford University Press, 2012). His research interests include the history and theory of photography, surveillance studies, and sport studies.

Rebecca Gowland is a professor of human bioarchaeology in the Department of Archaeology at Durham University. Her research focuses on the interrelationships between the body and society in the past, and she is particularly interested in the life course and age as aspects of social identity. Her publications include the coauthored book *Human Identity and Identification* (Cambridge University Press, 2013) and the coedited books *Care in the Past: Archaeological and Interdisciplinary Perspectives* (Oxbow, 2017) and *The Infant-Mother Nexus in Anthropology* (Springer, 2019). She also publishes widely in peer-reviewed journals on methodological and theoretical approaches to the study of skeletal remains.

Liv Hausken is a doctor and professor of media studies in the Department of Media and Communication at the University of Oslo and head of the interdisciplinary research group Media Aesthetics. She seeks to engage the arts and the humanities in studies of everyday technologies and their significance in the production of knowledge and power. Relevant publications include "Photographic Passport Biometry" in *PUBLIC: Art/Culture/Ideas* (March 2020), "Media Aesthetics" in *Oxford Bibliographies* (March 2018), "The Media Aesthetics of Brain Imaging in Popular Science" in *Reasoning in Measurement* (Routledge, 2017), and "The Archival Promise of the Biometric Passport" in *Memory in Motion* (Amsterdam University Press, 2016).

Matt Houlbrook is professor of cultural history at the University of Birmingham, United Kingdom. He works on twentieth-century Britain, with particular interests in gender, sexualities and selfhood, the politics of cultural life, and the politics and practice of writing cultural history. He is the author of *Queer London: Perils and Pleasures in the Sexual Metropolis, 1918–57* (University of Chicago Press, 2005) and *Prince of Tricksters: The Incredible True Story of Netley Lucas, Gentleman Crook* (University of Chicago Press, 2016). He is currently working on a global history of self-improvement, focusing on the rise and fall of the Pelman Institute and the correspondence course known as Pelmanism.

Rob Lederer is a lecturer in academic writing at the New College of the Humanities, and teaches English literature at Arcadia University. Both are in London.

Melissa M. Littlefield is a professor in the Department of English at the University of Illinois, Urbana-Champaign, and an affiliate of the Beckman Institute for Advanced Science and Technology. She is the author of *The Lying Brain: Lie Detection in Science and Science Fiction* (University of Michigan Press, 2011) and *Instrumental Intimacy: EEG Wearables and Neuroscientific Control* (Johns Hopkins University Press, 2018) and the coeditor of *The Neuroscientific Turn: Transdisciplinarity in the Age of the Brain* (University of Michigan Press, 2012). She is the coeditor of *Configurations: A Journal of Literature, Science and Technology*.

Andrew Mangham is professor of Victorian literature and medical humanities at the University of Reading. He is the author of *The Science of Starving* (Oxford University Press, 2020), *Dickens's Forensic Realism* (Ohio State University Press, 2016), and *Violent Women and Sensation Fiction* (Palgrave Macmillan, 2007). He has also edited a number of essay collections, including *Literature and Medicine in the Nineteenth Century* (Cambridge University Press, 2020) and *The Cambridge Companion to Sensation Fiction* (Cambridge University Press, 2013).

James Purdon is a lecturer in modern and contemporary literature at the University of St. Andrews. He is the author of *Modernist Informatics: Literature, Information, and the State* (Oxford University Press, 2016) and the editor of

British Literature in Transition 1900–1920: A New Age? (Cambridge University Press, 2021).

Victoria Stewart is reader in modern and contemporary literature at the University of Leicester and has published widely on World War II writing, Holocaust writing, and crime writing, with books including *Crime Writing in Interwar Britain: Fact and Fiction in the Golden Age* (Cambridge University Press, 2017). She is currently working on representations of Nazi war-crimes trials in 1940s and 1950s Britain.

Tim Thompson is a professor of applied biological anthropology and associate dean (academic) at Teesside University. A fellow of the Chartered Society of Forensic Sciences and an honorary fellow of the faculty of Forensic and Legal Medicine, his research focuses on the human body and how it changes, the recording and visualization of forensic evidence, and the practice of forensic anthropology. His publications include the coauthored *Human Identity & Identification* (Cambridge University Press, 2013) and the edited *Human Remains—Another Dimension: The Application of Imaging to the Study of Human Remains* (Elsevier, 2017) and *The Archaeology of Cremation* (Oxbow, 2015).

Index

Aas, Katja Franko, 187
ABC Correspondence Schools, 10, 25, 27, 29, 35
About, Ilsen, 7
Altick, Richard D., 74, 76
amnesia, 54–55, 189–98
Ancestry.com, 124–26, 129, 135
anonymity, 5, 183–84, 187, 190, 218, 223–24
anthropometry, 8, 25, 50
Appleby, Timothy, 141
Armstrong, Nancy, 74
As for the Woman (Iles), 82, 88, 92–94
Ashbourn, Julian, 162, 169
athletics, 109
Aus, Jonathan P., 160

Ballard, J. G., 16, 202–9, 212–19
Banville, John, 82
Barker, Nicola, 16, 202–3, 207–19
Barnard-Wills, David, 185
Batchen, Geoffrey, 147
Baudrillard, Jean, 191, 238
Bauman, Zygmunt, 134, 228–29
Beckman, Frida, 204
Before the Fact (Iles), 82, 84, 87–92
Bell, Charles, 25
Bender, John, 14
Benjamin, Walter, 232
Bentham, Jeremy, 226
Bertillon, Alphonse, 25
Big Clock, The (Fearing), 11, 57–62
biometrics, 9, 14, 186–87
  facial recognition technology (FRT), 14, 159–75
  iris scanning, 186–87
biotechnology, 206–9, 217
  and genetics, 122–35
Birchall, Clare, 188, 212, 238

"Black Box" (Egan), 237
Black Curtain, The (Woolrich), 52–54, 62
Blackford, Katherine, 29–33, 36, 37
Blair, Tony, 184
Blayney, Steffan, 26
Bleak House (Dickens), 71
Blunderer, The (Highsmith), 63
Bogard, William, 191
Bolnick, Deborah A., 125
Bones (TV series), 105
Bourne, Jason (film series) 15–16, 183, 189–98
Bouw, Matthijs, 214
Bowen, John, 68
Breckenridge, Keith, 7
Brown, James, 7
Brown, Michelle, 144–45

Caplan, Jane, 4, 6, 8, 12, 167
Carby, Hazel, 127
Carnegie, Dale, 28
Carrabine, Eamonn, 144–45
Central Intelligence Agency (CIA), 190–98
Chandler, Raymond, 11, 51, 52, 63
  Big Sleep, The, 50
  "Simple Art of Murder, The," 51, 56
Chin, Frank, 127
Christie, Agatha, 82
Chu, Patricia E., 6
citizenship, 6, 40, 128, 211–13, 218
Civil Rights Movement, 122
clairvoyance, 33
Cole, Simon A., 7
Collins, Philip, 79
Comeau, Marie France, 140, 142
Cooper, Melinda S., 122–23, 132
Corbusier, Le, 205
CSI (TV series), 7, 105, 111, 146

Daily Sketch, 23
Dangerfield, Fred, 34
Darwin, Charles, 25, 73
Davis, Angela Y., 134
Davis, Lennard J., 121
Davis, Natalie Zemon, 5
Deleuze, Gilles, 188
Derrida, Jacques, 232–33
detective fiction (classic), 85–87
Dickens, Charles, 11, 67
   Bleak House, 71
   Household Words, 75
   Martin Chuzzlewit, 76–78
   Mystery of Edwin Drood, The, 68–70, 76
   "Night Walks," 67, 68, 70, 78
Dijk, Jose van, 147
DNA
   databases, 185–86
   profiling 8, 12, 110–11, 185–87
Dostoyevsky, Fyodor, 84
Doyle, Richard, 135
Dupré, John, 124
Dussere, Erik, 59

Egan, Jennifer, 16, 223–40
   "Black Box," 237
   Look at Me, 230–31
   Visit from the Goon Squad, A, 223–40
Elliott, Kamilla, 7n24
Epstein, Debbie, 191
Ericson, Richard V., 188, 223

Facebook, 3. *See also* Zuckerberg, Mark
Farrow, John, 61
Fausto-Sterling, Anne, 107
Fearing, Kenneth, 11, 63
   Big Clock, The, 57–62
   "Escape," 55–57
Federal Bureau of Investigation (FBI), 49–50
Feldman, Robin Cooper, 169
Ferguson, Andrew Guthrie, 224
fingerprints, 8, 12, 33, 50, 175, 186–87
Fixer Chao (Ong), 128
Flickr, 148–49
forensic anthropology, 101–16
forensic science
   Daubert criteria, the, 106, 115
   medical jurisprudence, 11, 65–67, 71
   and photography, 71–75

Foth, Marcus, 203
Foucault, Michel, 72, 191, 211–12, 226
Frederick, Pauline, 35
French Revolution, the, 6
Frosh, Paul, 147
Frye, Alexis Everett, 128, 133, 135

Galton, Francis, 11, 12, 71–74, 77
Gąsiorek, Andrzej, 206n20
Gates, Henry Louis, Jr., 121
Gates, Kelly A., 5, 6, 162, 173–74
Geary, Patrick J., 125
Gellen, Kata, 8
gender (politics), 93, 133–34
   gendered violence, 140–43
   in relation to sex, 107–9
genetics
   and biotechnology, 13, 122–35
   epigenetics, 113–14
   Human Genome Project, 121, 124, 127
   and race, 121–35
Gerlach, Neil, 126
Giroux, Henry, 134
Globe and Mail (Canada), 139, 141, 146
Google, 224
Gordon, Avery, 233n31
Graves, Robert, 27, 36
Green, David, 144
Groebner, Valentin, 5, 14
Guattari, Felix, 188
Gunnarsdóttir, Kristrún, 159, 174

Hacking, Ian, 124–25
Haggerty, Kevin D., 188, 223
Hall, Stuart, 5
Hammett, Dashiell, 11, 49, 51
   Red Harvest, 49
Hand, Martin, 147, 149
H(A)PPY (Barker), 16, 202–3, 207–19
Haraway, Donna, 133
Harper, Frances Watkins, 126–27
Hayward, Keith, 145, 149
Higgs, Edward, 5, 7, 8, 9, 185–87
Highsmith, Patricia, 12, 63, 84
Hitchcock, Alfred, 89
Hodge, Alan, 27, 36
Horsley, Lee, 54
Howard, Norman, 34
Human Stain, The (Roth), 124
Humble, Nicola, 86

identity cards, 50
  Identity Cards Act 2006 (UK), 184–85
  National Registration Identity Cards (UK), 183–84
Iles, Francis, 12
  As for the Woman, 82, 88, 92–94
  Before the Fact, 82, 84, 87–92
  Malice Aforethought, 82–83, 85, 87–91, 94–95
impersonation, 1–4
Instagram, 149
International Civil Aviation Organization, *Doc 9303*, 14, 159–75
Introna, Lucas D., 161–62, 171–73
Iola Leroy (Harper), 126–27
Iser, Wolfgang, 235

Jain, Anil K., 175
Jeffreys, Alec, 110, 186
Jesse, F. Tennyson, 92
Jewell, Matthew, 202–3, 213
Johnston, Katherine D., 224n5, 226n14, 229
Jordan, Tim, 211
Joy (magazine), 23–36 passim
Joyce, Patrick, 209
Joyce, Rosemary, 103
Judging Character at Sight (Blackford), 30–32

Killer Inside Me, The (Thompson), 63
Kingston, Maxine Hong, 128
Klein, Norman M., 237
Komninos, Nicos, 204

Lalvani, Suren, 144
Lange, Michiel de, 214
Latour, Bruno, 143
Law, Frederick Houk, 37
Lee, Rachel C., 132
Leopold, Nathan, 91
Lister, Martin, 147
Lloyd, Jessica, 140–42
Locard, Edmond, 4
Loeb, Richard, 91
Lombroso, Cesare, 71, 144
London Metropolitan Police, 1
Lonergan, Gayle, 7
Look at Me (Egan), 230–31
Lynch, Michael, 143
Lyon, David, 14, 187–89, 224, 228

Macherey, Pierre, 73
Machin, David, 147
Malice Aforethought (Iles), 82–83, 85, 87–91, 94–95
Manovich, Lev, 147, 236–37
Martin Chuzzlewit (Dickens), 76–78
Meloni, Maurizio, 114
Meskell, Lynn, 103
Millais, John Everett, 75
Milne, A. A., 51
Mishra, Pankaj, 237
Mitchell, W. J. T., 14, 143, 147
Mnookin, Jennifer, 144
modernism, 54
Moling, Martin, 233n30, 234n33
Morozov, Evgeny, 216
Morris, David B., 121
Mulhall, Stephen, 192
Murray, Susan, 140n3, 147, 149
Mutlu, Can E., 159, 169
Mystery of Edwin Drood, The (Dickens), 68–70
My Year of Meats (Ozeki), 13, 121–23, 127–35

Naafs, Saskia, 204n9
Nakhaeizadeh, Sherry, 112
National Security Agency (NSA), 225
Nelson, Alondra, 217
neoliberalism, 134, 218
New Kind of Monster, A (Appleby), 141
Nietzsche, Friedrich, 91
"Night Walks" (Dickens), 67, 68, 70, 78
Nissenbaum, Helen, 161–62, 171–73
noir thriller, 10–11, 49–63
No-No Boy (Okada), 127–28
Novello, Ivor, 23–40 passim

O'Hagan, Andrew, 1–4, 9
Okada, John, 127–28
Ong, Han, 128
O'Rourke, Chris, 34, 35
Ostrowidzki, Eric A., 218
Ozeki, Ruth, 13, 121–23, 127–35

Palmer, William, 92
Panek, LeRoy Lad, 89
panopticism, 190–91, 210, 212, 226–27
Parenti, Christian, 189
Pasquale, Frank, 209

passports, 50, 55
   and biometrics, 159–75
   and border control, 159–75
Peirce, Charles Sanders, 8
Pelman Institute, the, 10, 25, 26, 29, 32, 35, 41
People (London), 183
photography, 14
   Abu Ghraib images, 146–48
   character competitions, 24–25, 28, 29
   and criminality, 139–50
   and facial recognition technology (FRT), 165–75
   in forensic science, 71–75
   mug shots, 25, 33
   and self-improvement, 23–42
physiognomy, 25, 30, 33, 71–72
Picon, Antoine, 202n2
Picturegoer (UK), 31, 34, 38
Pierres, Stella, 23–24, 27–28, 37
Plain, Gill, 85
Ploeg, Irma van der, 186–87
Poe, Edgar Allan, 72, 75, 78
Poster, Mark, 187, 210, 226, 230–33, 239
Prabhakar, Salil, 175
psychoanalysis, 27, 53, 83, 86–88, 196
psychology
   criminal, 85–87
   popular, 26, 41

Rabinowitz, Paula, 54
race, 12, 13
   colonialism, 31
   eugenics, 72
   and genetics, 121–35
   racial profiling, 216
   and technology, 217
Rafter, Nicole Hahn, 144
Rattenbury, Alma, 94
Ratti, Carlo, 204
Red Harvest (Hammett), 49
Rich, Irene, 35
Richardson, Sarah S., 123
Ricoeur, Paul, 161, 167
Robertson, Craig, 7, 55
Rommetveit, Kjetil, 159, 174
Rosen, David, 14
Ross, Arun, 175
Roth, Philip, 124
Rowling, J. K., 82
Rozelle, Lee, 205
Rubenstein, Daniel, 147, 149

S and Marper v. United Kingdom (2006), 185–86
Sadowski, Jathan, 209
Sala, George Augustus, 68
Saler, Michael, 24
Salter, Mark B., 159, 169, 175
Santesso, Aaron, 14
Sayers, Dorothy L., 85
Schaffers, Hans, 204
Schine, Cathleen, 237
Science, 112
Scott, Justice Robert, 142
"Secret Lives of Ronal Pinn, The" (O'Hagan), 1–4, 9
Sekula, Allan, 144
self-improvement, 23–42
semiotics, 8–9
Sherwin, Richard K., 144
Shilling, Chris, 103
"Simple Art of Murder, The" (Chandler), 51, 56
Sluis, Katrina, 147, 149
smart cities, 202–19
Smiles, Samuel, 40
Smyth, Jim, 146
Snowden, Edward, 188, 225
social security number (US), 55
Sola, Vincent de, 34, 37
Sontag, Susan, 146–47, 150
Standard Art Company, 26, 30
Steinberg, Deborah Lynn, 191
Stevens, Hallam, 123
Stoner, George, 94
Suckling, Meagan, 143
Sunday News (London), 24, 29–33 passim
Sunday Times (London), 82–83
Super-Cannes (Ballard), 16, 202–9, 212–19
Surveillance, 15, 16, 186–98
   biosurveillance, 206–9
   CCTV, 15, 186, 190–91, 213
   data doubles, 16, 61
   digital databases, 16, 223–25, 230–40
Suspicion (Hitchcock), 89
Swanson, Gloria, 37
Szreter, Simon, 7

Tagg, John, 144
Talented Mr. Ripley, The (Highsmith), 63
Taylor, Alfred Swaine, 11, 65, 66, 67
Thomas, Ronald R., 7, 72, 74
Thompson, Jim, 63

Thompson, Laura, 82
Thomson, H. Douglas, 86
Thomson, Matthew, 26
Times (London), 65, 66
Timson, Judith, 146
Todorov, Tsvetan, 84
Torpey, John, 4, 6, 14
Townsend, Anthony M., 202n1
Tripmaster Monkey (Kingston), 128
Turnbull, Malcolm, 83
Turner, Bryan S., 103
Twitter, 3

Umphrey, Martha Merrill, 144
"Unwritten Novel, An" (Woolf), 39–41
US Enhanced Border Security and Visa Entry Reform Act of 2002, 160

Veel, Kristin, 236
Visit from the Goon Squad, A (Egan), 16, 223–40

Waal, Martijn de, 214
Walker, Phillip L., 108
watch lists, 161–62, 170–71
wearable technology, 208–10
Willcock v. Muckle, 183–84
Williams, Russell, 13, 139–43, 145–46, 148–50
Woolf, Virginia, 39–41
Woolgar, Steve, 143
Woolrich, Cornell, 11, 57, 63
    Black Curtain, The, 52–54, 62
Wright, Alexa, 72

Zamyatin, Yevgeny, 205
Zappen, James, 235
Zuboff, Shoshana, 224–25, 228, 237–38
Zuckerberg, Mark, 197. *See also* Facebook

Lightning Source UK Ltd.
Milton Keynes UK
UKHW041104270222
399275UK00002B/57